W0107112

Risk Management of Chemicals in the Environment

NATO • Challenges of Modern Society

A series of edited volumes comprising multifaceted studies of contemporary problems facing our society, assembled in cooperation with NATO Committee on the Challenges of Modern Society.

Risk Management of Chemicals in the Environment

Edited by

Hans M. Seip

and

Anders B. Heiberg

Center for Industrial Research
Oslo, Norway

Published in cooperation with
NATO Committee on the Challenges of Modern Society

PLENUM PRESS • NEW YORK AND LONDON

Library of Congress Cataloging in Publication Data

Risk management of chemicals in the environment.

(NATO challenges of modern society; v. 12)
"Published in cooperation with NATO Committee on the Challenges of Modern Society."
Outcome of a pilot study, initiated in Oslo, Norway in 1984.
Bibliography: p.
Includes index.
1. Toxicity testing. 2. Pollution. 3. Health risk assessment. 4. Risk management. I. Seip, Hans Martin, 1937–   . II. Heiberg, Anders B. III. North Atlantic Treaty Organization. Committee on the Challenges of Modern Society. IV. Series.
RA1199.R56   1988   363.1'79   88-31626
ISBN-13: 978-1-4684-5606-6   e-ISBN-13: 978-1-4684-5604-2
DOI: 10.1007/978-1-4684-5604-2

Proceedings of the NATO Committee on the Challenges of
Modern Society Pilot Study on Risk Management of Chemicals
in the Environment, initiated with Norway
in April 1984

© 1989 Plenum Press, New York
Softcover reprint of the hardcover 1st edition 1989

A Division of Plenum Publishing Corporation
233 Spring Street, New York, N.Y. 10013

All rights reserved

No part of this book may be reproduced, stored in a retrieval system, or transmitted in any form or by any means, electronic, mechanical, photocopying, microfilming, recording, or otherwise, without written permission from the Publisher

PREFACE

The Council of the North Atlantic Treaty Organization (NATO) established the "Committee on the Challenges of Modern Society" (CCMS) in 1969. The CCMS was charged with developing meaningful environmental and social programmes for solving existing problems and developing long-range goals for environmental protection.

In 1983, at the Fall Plenary of the CCMS, the Norwegian delegate Dr. H.C. Christensen, proposed a Pilot Study on "Risk Management of Chemicals in the Environment". A draft proposal, written by Dr. Kari Kveseth of the Center for Industrial Research in Oslo, was presented. Dr. Christensen also informed the participants at the meeting that Norway was willing to act as the Pilot Country. The project was initially planned for 3 years, but it was later extended through 1987.

The inaugural meeting was held in Oslo in April 1984 with participants from Denmark, Greece, France and Italy, in addition to representatives from several Norwegian institutions. The attendees concluded that a Pilot Study, as delineated in the draft proposal, would be useful, and it was decided to work out a detailed project plan based on the proposal.

The Pilot Study activity was rather low in 1984, but slowly the study gained momentum. This was due in part to active contributions from scientists receiving CCMS fellowships for working on the project. In 1985 the Netherlands, represented by the TNO, and the USA, represented by the U.S. Environmental Protection Agency, joined the Pilot Study. Pilot Study meetings were arranged in Oslo (March 1985), Washington (November 1985), Rome (April 1986), Liège, Belgium (November 1986) and in Oslo (October 1987).

Although the Pilot Study was initially aimed at risk management, it soon became apparent that it was necessary to include aspects of risk assessment as well. Of course, one could not hope to cover all aspects of risk assessment and risk management in a study of this kind. The participants were not chosen with such a goal in mind. Nevertheless, many important topics were covered. In general, the participants contributed projects in which they were involved when the Pilot Study was initiated. As a result, each individual subproject in the study was conducted mainly as a national undertaking. However, Pilot Study meetings and visits between Study Group members had in some cases considerable influence on the project development.

The outcome of the Pilot Study is reported in this book. It should be emphasized that the views expressed in the various chapters do not necessarily represent the official views of the participating countries. Members of the Study Group and persons with CCMS fellowships are listed in Appendix A and B, respectively.

Conclusions and recommendations to the CCMS were presented in a Summary Report which was submitted to the CCMS at the committee's Spring Plenary in 1988. The Summary Report is reproduced in Appendix G.

The Editors wish to express their warmest thanks to all those who have participated in the Pilot Study. We also thank Judy B. Theisen at the EPA Technical Information Staff who has corrected the language in many of the chapters. Financial support from the Norwegian Ministry of Environment and from the Norwegian Council for Scientific and Industrial Research is gratefully acknowledged.

Hans M. Seip
Pilot Study Director

Anders B. Heiberg
Pilot Study Officer

June, 1988

CONTENTS

CHAPTER 1: PILOT STUDY OF RISK MANAGEMENT OF CHEMICALS
IN THE ENVIRONMENT: AN INTRODUCTION

H.M. Seip and A. Heiberg

CHAPTER 2: QUANTIFICATION OF HEALTH RISK DUE TO CHEMICALS:
METHODS AND UNCERTAINTIES

D.H. Trønnes and A. Heiberg

CHAPTER 3: METHODS FOR ASSESSING THE RISK OF
ENVIRONMENTAL CONTAMINATION

C.L. van Deelen

CHAPTER 4: METHODS USED IN THE UNITED STATES FOR THE
ASSESSMENT AND MANAGEMENT OF HEALTH RISK
DUE TO CHEMICALS

J.W. Falco and R.V. Moraski

CHAPTER 5: ASSESSMENT OF ECOLOGIC RISKS RELATED TO CHEMICAL
EXPOSURE: METHODS AND STRATEGIES USED IN THE
UNITED STATES

J.W. Falco and R.V. Moraski

CHAPTER 6: ATTITUDES TOWARD RISK-BENEFIT ANALYSIS FOR MANAGING EFFECTS OF CHEMICAL EXPOSURES

J.L. Regens

CHAPTER 7: MODELLING BEHAVIOUR OF POLLUTANTS IN SOIL FOR RISK ASSESSMENT PURPOSES

C. Lupi

CHAPTER 8: ENVIRONMENTAL AND HEALTH IMPACT
ASSESSMENT OF SOIL POLLUTANTS.
THE SEVESO ACCIDENT AS A TYPICAL EXAMPLE

G.A. Zapponi and C. Lupi

CHAPTER 9: MATHEMATICAL AND BIOLOGICAL UNCERTAINTIES
IN THE ASSESSMENT OF A PERMISSIBLE
BLOOD LEAD CONCENTRATION

F. Sartor and D. Rondia

CHAPTER 10: USE OF FORMAL METHODS IN EVALUATING
COUNTERMEASURES TO COASTAL WATER POLLUTION.
A CASE STUDY OF THE KRISTIANSAND FJORD,
SOUTHERN NORWAY

A. Heiberg and K.-G. Hem

CHAPTER 11: ABATEMENT OF AIR POLLUTION IN OSLO

D.H. Trønnes

CHAPTER 12: SUMMARY, CONCLUSIONS AND RECOMMENDATIONS

A. Heiberg and H.M. Seip

PILOT STUDY ON RISK MANAGEMENT OF CHEMICALS IN THE ENVIRONMENT:

AN INTRODUCTION

H. M. Seip and A. Heiberg

Center for Industrial Research
Oslo, Norway

1   BACKGROUND OF THE PILOT STUDY

Assessment and management of risks connected to health and safety protection have very old origins. In fact, procedures for regulating the consumption of potentially dangerous foods are mentioned in the Bible. Laws including some risk management principles were also present in legislation from the Roman Empire and Middle Ages. English laws were issued in the thirteenth and fourteenth century for the prevention of health risks by unwholesome or damaged foods. As early as 1661, John Evelyn discussed health problems by smog in London. In 1855 the question of how to evaluate a clean environment was expressed by Chief Seattle in his famous letter to the American President, in which he states : "If we do not own the freshness of the air and the sparkle of the water, how can you buy them?"

Despite these early attempts at risk assessment and risk management, it is only during the last decades that such procedures have become more generally recognized as useful tools in decision making at local, regional and national levels. At present, in North America, in most European countries and in many other countries governmental agencies and scientific institutions use official and well established procedures to characterize and provide information useful to decision optimization. This approach has, for example, led to the procedure for carcinogenic risk assessment which is presently used in the USA and in other countries, and which has been proposed as reference method also by the World Health Organization.

In spite of public awareness and some serious attempts to reduce environmental pollution, the problems seem to increase on a global scale. We have recently witnessed catastrophes such as the accident in the nuclear reactor in Chernobyl, the spread of deadly methyl isocyanate from the plant in Bhopal, and serious spills of chemicals in the Rhine River. In these cases the cause of the damage is obvious and the effect immediate. In other cases the adverse effects of chemicals may become apparent first after a long period of time, as illustrated by the recent forest damage in Central Europe and the decrease in stratospheric ozone in the Antarctic. Under such circumstances it may be very difficult to establish casual relationships with certainty. Although it seems to be accepted that pollutants play an important role in the forest damage, there is considerable argument with respect to the relative effects of various pollutants and to the role of natural causes (e.g. frost, drought).

The World Commission on Environment and Development, which was established by the United Nations General Assembly, described many aspects related to chemicals in the environment in the recent report entitled "Our Common Future" (WCED, 1987). The commission proposes a number of legal principles for environmental protection and sustainable development, e.g.:

- States shall establish adequate environmental protection standards and monitor changes in and publish relevant data on environmental quality and resource use.

- States shall make or require prior environmental assessments of proposed activities which may significantly affect the environment or use of a natural resource.

Although basic concepts of risk assessment and risk management have been applied in past cultures, they were not formally defined or accompanied by detailed technical and scientific methodologies. However, it is increasingly recognized that formal methods may be helpful in making decisions on problems concerning health or environment. The decision-makers are often in a very difficult situation. On the basis of a huge amount of data, which may be interpreted in conclicting ways by experts, they must try to reach rational decisions on how to protect health and environment. Although the formal decision-analytic methods available still have considerable weaknesses, they force the decision-maker to structure the problem, and, as far as possible, to quantify the consequences of various options. The structuring process also provides valuable opportunities for communication between the decision-maker and various interest groups.

Even though considerable progress has been made over the last few decades in the development of methods of risk assessment and risk management, decision making in the environmental area still tends to be an exceedingly difficult task. There is a pressing need for continued research and development in this area and for more experience with the use of decision-analytic tools in the management of chemical risk. The realization of this fact constituted the basis for the present Pilot Study on risk management of chemicals in the environment. The draft study proposal, presented to the CCMS at the committee's Fall Plenary 1983, stated:

Decision analysis as a tool is aimed at two levels. In order to ensure a more effective use of natural resources and set priorities among proposed countermeasures, it is important to develop a tool which is generally accepted by the national management. However, environmental risk management and several of the environmental problems discussed are of international concern and interest. Internationally there is an increasing activity in this field. Several methods for decision analysis/environmental risk management are being developed. In order to discuss environmental problems in international fora, we must aim at developing methods and criteria for risk management that we generally have agreed upon....

The Pilot Study on Environmental Risk Management shall give a review of existing methods for decision analysis, give recommendations and develop methods that may provide decision-makers with a better analytical tool in evaluating complex decision alternatives.

The objectives of the study were:

- To review and develop methods through studies of specific examples related to chemicals in the environment;

- To identify areas for further research and development; and

- To establish methods for risk assessment and risk management that may be used or adopted by member countries.

Although the Pilot Study was initially aimed primarily at risk management, it soon became apparent that it was also neccesary to include risk assessment to some extent. Of course, one could not hope to cover all aspects of risk assessment and risk management in a study of this kind. The participants were not chosen with that goal in mind. Nevertheless, many important subjects have been covered. In general, the participants contributed projects in which they were involved when the Pilot Study was initiated. As a result, the individual subprojects in the study were conducted mainly as national undertakings. However, Pilot Study meetings and visits between Study Group members had in some cases a considerable influence on the project development.

In the next two sections of this chapter the conceptual framework of the Pilot Study is discussed: Section 2 introduces some key concepts of risk assessment and risk management; Section 3 summarizes some of the major decision-analytic tools available for the management of chemical risk. Section 4 addresses another issue of great importance in environmental decision making: that of public acceptabilty. Finally, in section 5 the outcome of the Pilot Study is summarized.

2   COMPONENTS OF RISK ASSESSMENT AND RISK MANAGEMENT

Unfortunately, the nomenclature in the field of risk assessment and risk management is not universal. Even the word "risk" is used with different meanings. The risk associated with technical equipment (e.g., a chemical plant) is often meant to include both the probability and the consequences of an accident. If the number of fatalities is the consequence considered and P_N is the probability of $\underline{N}$ fatalities, the risk may be expressed as

$$P_N \cdot N^q,$$

where q is a positive number.

It is often argued that one big accident with many fatalities is more serious than several smaller ones which together give the same number of fatalities, so that $q > 1$. However, it has also been stated that there is insufficient justification for using a simple function of N to model social impacts (Slovic et al., 1984).

In connection with human activities the term "risk" is normally meant to imply only the probability of a particular event. In this book "risk" will mostly be used in this sense. To illustrate the concept, some risks of death and the estimated uncertainties as given by Wilson and Crouch (1987), are presented in Table 1.1. Clearly, when considering risk-

Table 1.1 Mean Values with Uncertainty for Some Commonplace Risks of Death (From Wilson and Crouch, 1987)

Action	Annual risk	Uncertainty
Motor vehicle accident (total)	2.4×10^{-4}	10%
Motor vehicle accident (pedestrian only)	4.2×10^{-5}	10%
Home accidents	1.1×10^{-4}	5%
Electrocution	5.3×10^{-6}	5%
Air pollution, eastern United States	2×10^{-4}	Factor of 20 downward only
Cigarette smoking, one pack per day	3.6×10^{-3}	Factor of 3
Sea-level background radiation (except radon)	2×10^{-5}	Factor of 3
All cancers	2.8×10^{-3}	10%
Four tablespoons peanut butter per day	8×10^{-6}	Factor of 3
Drinking water with EPA limit of chloroform	6×10^{-7}	Factor of 10
Drinking water with EPA limit of trichloroethylene	2×10^{-9}	Factor of 10
Alcohol, light drinker	2×10^{-5}	Factor of 10
Police killed in line of duty (total)	2.2×10^{-4}	20%
Police killed in line of duty (by felons)	1.3×10^{-4}	10%
Frequent flying professor	5×10^{-5}	50%
Mountaineering (mountaineers)	6×10^{-4}	50%

reducing measures the decision-maker also has to take into account how many people are exposed to the particular risk in question.

In discussing the risks caused by chemicals it is common to distinguish between risk assessment, which in principle can be carried out objectively, and risk management, which involves preferences and attitudes and therefore is an essentially subjective activity (National Research Council, 1983; U.S. EPA, 1984).

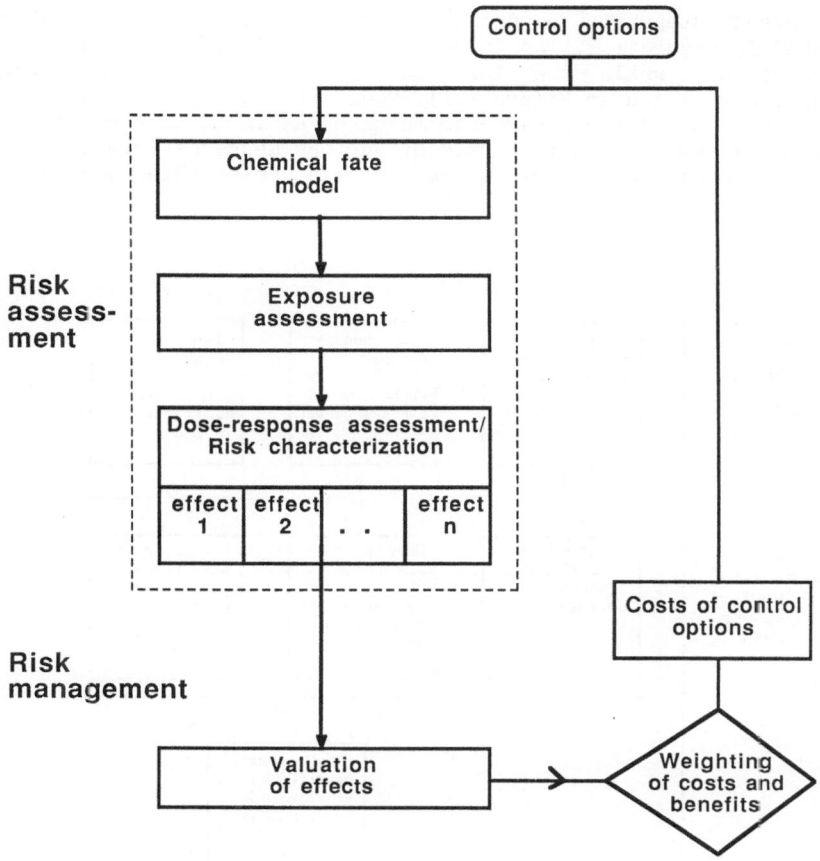

Figure 1.1 Elements of risk assessment and risk management (From U.S. EPA, 1984)

Risk assessment, as defined by the U.S. Environmental Protection Agency, among others, can be characterized by the key words:

- Hazard identification

- Dose-response assessment

- Exposure assessment

- Risk characterization

These terms are explained in Figure 1.1 (see also Chapter 4). Other organizations have developed slightly different approaches to risk assessment (see Chapter 3).

Risk management is defined as the process of weighing policy alternatives and selecting the most appropriate regulatory action, integrating the results of risk assessment with engineering data and with social, economic and political concerns to reach a decision. The term "decision analysis" may be used to cover both risk assessment and risk management.

Important steps in a decision analysis of pollution control are illustrated in Figure 1.2. The upper part of the left hand side involves risk assessment, while the lower part and the right hand side involves risk management (see also Figure 1.1). Models for the dispersion and fate of pollutants and for exposure are required as input in a response model for effects. In most cases the costs of various measures (right hand side of the figure) can be estimated with reasonable accuracy. The

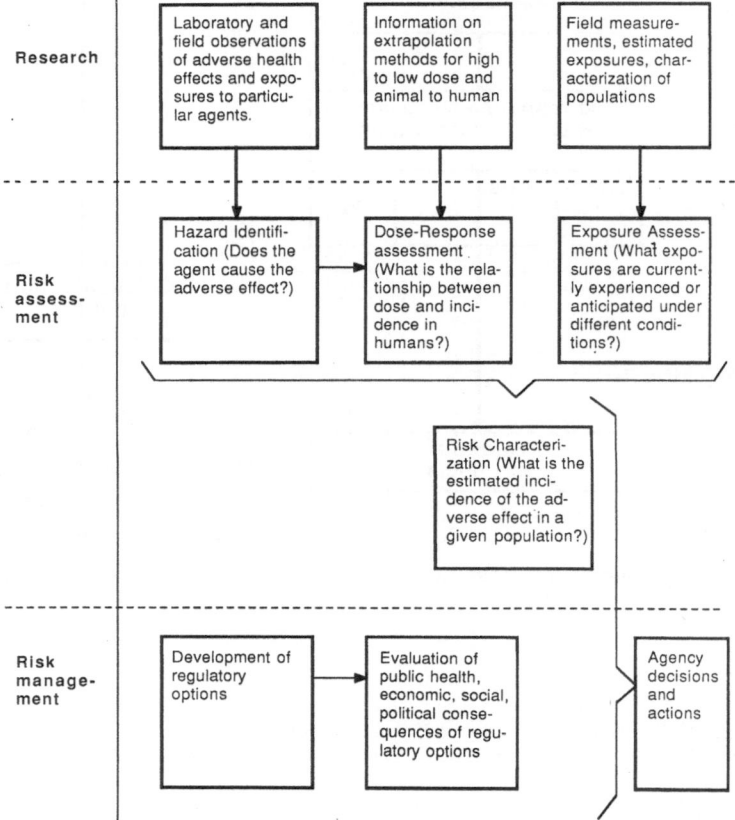

Figure 1.2 Components of an analysis of pollution control

uncertainties are generally much larger in the estimates of exposure and, in particular, in the relationships between dose and response (e.g., the per cent mortality). It is of utmost importance that the uncertainties originating from all steps in the analysis are included in the presentation of the final results (see, for example, Suter et al., 1987).

Though the uncertainties often are large, both exposure doses and related effects can, in principle, be determined objectively, as already noted. This is not the case for the valuation of effects which is necessarily subjective. We are often dealing with commodities for which there is no market price. There is, for instance, no absolute answer to the question of the value of human life (see, for example, Føllesdal, 1986) or that of an antique building. And even if market prices do exist, their usefulness may be questioned. The problems were elegantly illustrated by Oscar Wilde when he defined a cynic as a man who knows the price of everything and the value of nothing.

Despite the problems associated with the pricing of human life, it is interesting to note that in the Netherlands acception criteria for individual and group risks resulting from industrial activities, including transport, have been developed. These criteria were first used for liquefied petroleum gas (LPG) safety matters.

3   QUANTITATIVE DECISION ANALYSIS TECHNIQUES

Detailed discussions of various techniques for decision analysis, suited for the study of environmental policies, is beyond the scope of this chapter. We will use "decision analysis" as a general term for a series of methods. Others use this term only for techniques designed to take uncertainties explicitly into account. Some key methods in decision analysis are mentioned below. For further details, see Morgan (1983).

3.1   Cost-Benefit Analysis

Cost-benefit analysis is a procedure for determining whether the expected benefits from a proposed action outweigh the expected costs. It is necessary that all costs and benefits be expressed in monetary units. For some benefits market prices may be obtained. In other cases it may be possible to find a "surrogate market." Alternatively, people's willingness to pay may be obtained from interviews.

3.2   Contingent Cost-Benefit Analysis

In contingent cost-benefit analysis unit prices are estimated for some of the benefits as in a normal cost-benefit analysis. For those benefits that are more difficult to express in monetary terms, the unit prices are varied to find the smallest values that make the net benefit positive. Alternatively, one may find the range in monetary units that makes a certain control option the best one. This procedure does not solve the difficult problems of valuating commodities for which there is no market price, but it is at least not necessary to attach a definite value to, for example, human life.

3.3   Cost-Effectiveness Analysis

Cost-effectiveness analysis is an examination of the costs of alternative means to achieve a given goal. The goal is usually expressed in non-monetary units. For example, it may be specfied as a certain pollution level.

3.4   Risk-Benefit Analysis

In a risk-benefit analysis no attempt is made to measure all consequences of the activity of concern in commensurate units. A monetary estimate is obtained for the net economic benefit of the activity, which could be, for example, the production of a chemical that has useful applications but are toxic to humans. The other benefits associated with the activity, e.g., employment, are expressed in other, more natural units. The (non-economic) costs involved, such as the risk of adverse health effects due to the use of a chemical product, are also quantified in non-monetary units. Usually, the formal part of the analysis is ended at this stage. The final task of balancing the benefits against the risks is left to the decision-maker. Often, the conclusions drawn by the decision-maker will implicitly imply monetary values for the non-market commodities involved in the analysis.

3.5 Multi-Objective Decision Making Techniques

The term "multi-objective decision making techniques" is often reserved for techniques that are not variations of cost-benefit analysis; comparisons of benefits (in pollution control often reduced damage) are made without expressing them in monetary units. Although a number of such techniques have been developed (Zeleny, 1982), only a few will be discussed here.

Multiattribute utility analysis (MUA) may be derived from a set of axioms on how people make decisions when the outcome of the possible alternatives is uncertain. In an MUA costs and benefits are brought onto a common scale by assigning so-called "utilities" to the consequences considered. Thus, this technique makes it possible to balance in-commensurate consequences against each other without having to express everything in monetary terms. The utility of each consequence and the relative importance of the different types of consequences are derived from the answers given by individuals or groups to certain lottery questions. A special virtue of the MUA is that it allows one to take into account people's attitudes toward risk (Keeney and Raiffa, 1976). The Simplified Multiattribute Rating Technique (SMART), developed by Edwards (1977), is a simplified version of an MUA. An example of the use of this technique in environmental decision analysis is provided in Chapter 10 of this book.

The ELECTRE method, developed by Roy (1977), is an outranking method that is not based on a set of axioms. Roy and Bouyssou (1986), in comparing this method with the MUA, stated that ELECTRE is constructed to illuminate possible decisions by means of pragmatic ideas and intentional actions.

4 PUBLIC ACCEPTABILITY

A high degree of acceptance by the public is essential for the success of risk management decisions. Recent research has revealed a number of factors important for risk acceptance (Slovic, 1987). Voluntariness of exposure is one key mediator, but other factors, such as familiarity, control, catastrophic potential, equity, and level of knowledge also seem to play important roles. Sometimes experts have a tendency to ridicule the layperson's attitude toward risk. A much more constructive view is expressed by Slovic (1987):

"Perhaps the most important message from this [psychometric] research is that there is wisdom as well as error in the public attitudes and perceptions. Lay people sometimes lack certain information about hazards. However, their basic conceptualization of risk is much richer than that of the experts and reflects legitimate concerns that are typically omitted from expert risk assessments."

Fortunately, the importance of communication between technical experts, decision-makers and those affected by the decision is increasingly recognized. In a recent Royal Commission Report on Environmental Pollution (Royal Comm., 1984) it is stated that:

> "In a democracy it is an unhealthy sign when authority claims omniscience and dismisses grass-roots concerns as 'irrational'."

The chairman of the commission added in a recent paper (Southwood, 1985) that there are two major 'DON'Ts': Don't be secretive, and don't 'rubbish' the views of the 'concerned public'.

5   OUTCOME OF THE PILOT STUDY

The Pilot Study has resulted in useful cooperation among groups in the participating countries. The main outcome of the study is this book. The content of the book is briefly summarized here.

Chapters 2-6 deal to a large extent with methods for risk assessment and risk management, though care has been taken to illustrate the methods by practical examples. Discussions of dose-response relationships and related uncertainties for health effects are given in Chapter 2, whereas methods for assessing the risk of environmental contamination are the subject of Chapter 3. These topics are also addressed in Chapters 4 and 5, which deal with methods used in the United States for the assessment and management of health risk due to chemicals and with methods for assessment of ecological risks related to chemical exposure. Even if there are numerous difficulties in assessing dose-response relations for health effects, the problems are still much greater with respect to ecosystem responses to chemicals. As a result, guidelines for the assessment and management of ecological risk have not yet been as far developed in the United States as those applying to health risk. The views of EPA employees toward risk assessment, cost-benefit analysis, and the use of monetary values for human life are discussed in Chapter 6. At least for the limited sample of decision-makers considered, cost-benefit methods appears to be largely accepted. There is more scepticism against assigning monetary values to human life.

An example of a model to describe the fate of pollutants in soil is given in Chapter 7. Modelling of the fate of pollutants in water is briefly mentioned in Chapter 10.

Chapters 8 through 11 deal with specific cases of environmental risk. The first two are mainly on risk assessment: the Seveso accident in Italy (Chapter 8), and uncertainties in assessing acceptable concentrations of lead in blood, based primarily on Belgian data (Chapter 9). The latter two chapters are mainly on risk management: valuation of countermeasures to pollution in a Norwegian fjord (Chapter 10) and to air pollution in Oslo (Chapter 11).

In most cases decisions are arrived at without full use of formal methods, either because of limited resources or limited confidence in such methods. Simplified approaches are discussed in Chapters 3 and 11. The final chapter (12) gives conclusions and our recommendations to the CCMS.

In general, it should be possible to read any chapter independently of the others. To provide a more complete picture of the study, however, there are a number of cross-references to other chapters.

6   ACKNOWLEDGEMENTS

We are very grateful to Carmela Cavelli and Giovanni Alfredo Zapponi for their valuable contributions to this chapter.

7   REFERENCES

Edwards, W., 1977, Use of multiattribute measurement for social decision making, in: "Conflicting Objectives in Decisions," B.E. Bell, R.L. Keeney, and H. Raiffa, eds., Wiley, New York, pp. 247-276.

Føllesdal, D., 1986, Risk: philosophical and ethical aspects, in: "Risk and Reason. Risk Assessment in Relation to Environmental mutagens and Carcinogens," P. Oftedal and A. Brøgger, eds., Alan R. Liss, New York, pp. 41-52.

Keeney, R.L., and Raiffa, H.F., 1976, "Decisions with Multiple Objectives: Preferences and Value Trade-offs," John Wiley & Sons, New York.

Morgan, M.G., 1983, "The role of decision analysis in the implementation of environmental policies," Organisation for Economic Co-operation and Development, ENV/CHEM/CM 83.5, Annex, Geneva.

National Research Council of the National Academy of Sciences, 1983, "Risk Assessment in the Federal Government: Managing the Process," National Academy Press, Washington.

Roy, B., 1977, Partial preference analysis and decision aid : the fuzzy outranking relation concept, in: "Conflicting Objectives in Decisions," B.E. Bell, R.L. Keeney, and H. Raiffa, eds., Wiley, New York, pp. 40-75.

Roy, B., and Bouyssou, D., 1986, Comparison of two decision-aid models applied to a nuclear power plant siting example, Eur. J. Oper. Res., 25:200.

Royal Comm., 1984, Royal Commission on Environmental Pollution, tenth report: Tackling Pollution - Experience and Prospects, Her Majesty's Stationary Office, London.

Slovic, P., Lichtenstein, S., and Fishhoff, B., 1984, Modelling the social impact of fatal accidents, Manage. Su., 30:464.

Slovic, P., 1987, Perception of risk, Science, 236:280.

Southwood, T.R.E., 1985, The roles of proof and concern in the work of the Royal Commission on environmental pollution, Mar. Pollut. Bull., 16:346.

Suter, G.W., Barnthouse, L.W., and O'Neill, R.V., 1987, Treatment of risk in environmental impact assessment, Environ. Manage., 11:295.

U.S. EPA, 1984, Risk assessment and risk management: framework for decision making, EPA 600/9-85-002.

WCED, 1987, "Our Common Future," The World Commission on Environment and Development, Oxford University Press, Oxford.

Wilson, R., and Crouch, E.A.C., 1987, Risk assessment and comparisons: an introduction, Science, 236:267.

Zeleny, M., 1982, "Multiple-criteria Decision-Making," McGraw-Hill, New York.

QUANTIFICATION OF HEALTH RISK DUE TO CHEMICALS:

METHODS AND UNCERTAINTIES

D. H. Trønnes and A. Heiberg

Center for Industrial Research
Oslo, Norway

1   INTRODUCTION

As will be recalled from the previous chapter, quantification of the risk to human health caused by chemicals involves two separate tasks (see Figure 1.1). First, an assessment has to be made of the extent to which the population is exposed to the chemical considered. Second, the connection between the exposure level (dose rate and duration of exposure) and the probability of adverse health effects must be established. Before these two steps are begun, however, a hazard assessment is normally performed. That is, tests are conducted to determine whether or not the chemical poses a risk to human health or the environment. If the results of these tests show that the chemical is nontoxic or has an extremely small chance of getting in contact with the surroundings, a full risk assessment like that displayed in Figure 1.1 is unnecessary.

Health risks associated with chemicals have some characteristics which make the tasks mentioned above extremely complicated. First, ignorance with respect to the chemical's mechanism of action, including metabolism and pharmacokinetics, is rather common, making the determination of a dose-response curve for the effect in question difficult to perform. In particular, very little knowledge is available on the mechanisms that lead to cancer and other long-term health effects. Furthermore, the pathway linking the receptor of the chemical to the emission source is generally very complex, and the processes involved are often only partly understood. For these reasons performing an exposure assessment also tend to be a rather demanding task. Detecting adverse effects may also be difficult. Low-potency carcinogens, for example, may easily escape detection in tests aimed at revealing carcinogenicity.

These facts notwithstanding, some sort of risk assessment should be made when one is faced with a chemical that could pose a risk to human health. The possibility that the chemical may lead to catastrophic costs, although small, is by itself a sufficient argument for this view. Historically, there have been several epidemics on a limited scale due to the use of chemicals, including the thalidomide and asbestos cases. Also, accidental releases of chemicals have occurred with very severe consequences (see, for instance, Chapter 8 of this book, which deals with the Seveso accident). Such incidents further demonstrate the urgent need for chemical risk assessments.

In this chapter some key issues will be discussed relating to the quantification of the risk to humans caused by chemicals. The objective of the chapter, which is partly based on the work of Campbell et al. (1982), is to present the methods most widely used in quantitative chemical risk assessment, to describe the limitations of the various methods, and to discuss how the uncertainty associated with their use can be dealt with.

Emphasis in the present chapter is on dose-response assessment. This subject is discussed mostly at a general level. However, to illustrate the methods presented, a specific example of the determination of a dose-response curve is given (section 3.2). Also, in Chapter 9 some aspects of dose-response assessment associated with a specific health risk are discussed.

Exposure assessment is only briefly mentioned in this chapter. This does not mean that exposure assessment is of minor importance in chemical risk analyses. In fact, exposure is as important as toxicity in determining the overall risk. Methods of exposure assessment, however, tend to depend strongly on the particular case investigated. What can be said, in general, is that most methods involve the use of one or more mathematical models describing how the chemical is transported and transformed on its way from source to receptor. In Chapters 7 and 8 of this book, some aspects of exposure assessment related to soil pollution are addressed.

2   METHODS OF DETERMINING DOSE-RESPONSE RELATIONSHIPS

The goal of a dose-response assessment is to obtain a quantitative relationship between duration and level of exposure and the probability of contracting a certain adverse health effect. There are two main approaches to achieving this goal: epidemiological studies and laboratory testing on living organisms. The latter approach will be referred to here as the indirect method. Both approaches have strengths and weaknesses, and neither one alone is usually sufficient to establish a reliable dose-response relationship. In sections 2.1 and 2.2 the basic features of each approach are outlined.

2.1 Epidemiological Studies

Epidemiology is the study of the correspondence between incidence of a disease in a population and the environmental factors influencing that population. In one type of epidemiological study, the "cohort study", one follows a particular population and looks for differences in disease rates between groups differently exposed to a suspect substance. In another type, the "case-control study", one picks out, for each diseased individual (case), one or more disease-free persons (controls) in order to identify differences in environmental conditions. It has been demonstrated that the two types of studies are closely related (Prentice and Breslow, 1978), and a third proposed method, the "case-cohort study", can be viewed either as an inexpensive version of a cohort study or as an alternative to the case-control study using all information in the data. In a case-cohort study one follows a well-defined subpopulation, which is compared with the cases. Epidemiological data can be analyzed by means of various statistical methods, including analysis of categorical data, logistic regression, and survival analysis.

Dealing directly with effects in humans, epidemiology is generally considered the soundest basis for determining health risks caused by chemicals. The method has, nevertheless, severe limitations. Most notably, even when a study does indicate that a substance may have an adverse effect, it is generally too crude to enable a quantification of the health risk in question. Because of the lack of controlled exposure conditions, which is characteristic of epidemiological studies, one should include in the analysis all factors that could contribute significantly to the observed effect. Some of these factors may simply be unknown. Moreover, important information on those factors known to play a role, such as data on ambient concentrations of the substance investigated, may be scarce or completely unavailable.

Differences in disease rates among population groups, when occurring, are commonly quite small. Therefore, to be able to detect existing differences through epidemiological studies, the population samples must, in general, be large. This is another weakness of epidemiological studies.

Because of the above limitations, there are relatively few chemicals for which there is some epidemiological evidence of health damage in humans. As of 1985, 93 substances had been identified as carcinogens or probable carcinogens on the basis of epidemiological data (Dybing, 1986).

As a tool in risk management, epidemiology is also unsatisfactory in that it can provide evidence of adverse health effects only after years of exposure to a chemical, i.e., when damage is already a fact. Clearly, methods for dose-response assessment, devoid of the shortcomings of epidemiology, are needed. Indirect methods present one possibility.

2.2 Indirect Methods

Indirect (or experimental) methods are of two types: long-term tests on animals (animal bioassays) and short-term tests on microorganisms or cell cultures. In most cases the biological effect of concern is cancer.

2.2.1 Long-term Tests.
Long-term animal bioassays give the most direct evidence of carcinogenic potency of a chemical. An animal bioassay is a laboratory procedure in which scientists administer the test substance in one or more dose-levels to one or more groups of animals (often rodents) and compare their cancer incidence with that of a control group that has not been exposed to the substance.

In animal bioassays the scientist has full control over exposure conditions. This means that he or she can isolate the effect of a suspected carcinogen and reliably establish a dose-response relationship. Also, since the latency time of cancer is much shorter in animals than in humans, the animal's response to the administered substance can be observed within a reasonable period of time, usually within one or two years.

Despite its advantages, the practical usefulness of long-term tests in risk assessments is limited. Apart from the problems associated with extrapolating the results to humans, a subject which will be discussed in section 2.2.3, the main limitations are the long time required to perform the tests and the high costs involved. Thus, testing a single compound for carcinogenic activity may take up to three years and cost several hundred thousand dollars (Ames, 1979).

2.2.2 Short-term Tests. To overcome the resource requirements of animal testing, short-term in vitro tests for mutagenicity have been introduced. Most of these tests are designed to detect damage to DNA, the theory being that such damage is a major source of cancer and genetic birth defects.

The most widely known short-term test is the Ames test. In this test the mutagenic potency of a substance is measured by its ability to bring a mutant form of the Salmonella bacteria back to its normal form (Ames, 1979). In addition to the Ames test, there are several other short-term mutagenicity tests, including tests that use cultures of animal or human cells.

To be useful in cancer risk assessment, short-term mutagenicity tests should agree well with the results of animal tests for carcinogenicity. That is, if a chemical is positive in the short-term test, it should be positive in the animal test as well, and vice versa. Generally, the correlation is good but not perfect. The Ames test, for example, when applied to certain classes of substances, shows a high predictivity, meaning that a large fraction of the chemicals that are positive in the Ames test are also carcinogenic, but has a low sensitivity, i.e., the proportion of carcinogenic substances which are positive in the Ames test is relatively small. Thus, Resnick (1987), in a recent study of 73 chemical compounds, 44 of which were carcinogens, found a predictivity of 84% and a sensitivity of 45% in the Ames test. The question of which short-term test or battery of tests is best suited for predicting chemical carcinogenicity is a subject of continued research.

2.2.3 Extrapolating from Animals to Humans. In general, the evidence for carcinogenicity obtained from animal tests agrees well with the results of epidemiological studies. Thus, of the 93 chemicals previously referred to, i.e., those for which there was some epidemiological evidence of human carcinogenicity, all show sufficient evidence for carcinogenicity in animal experiments (Dybing, 1986). This suggests that animal bioassays may be helpful in assessing cancer risk to humans. To relate the results from animal experiments to humans is, however, far from straightforward. In order to secure response in a sufficiently large fraction of the test population, animal experiments are usually carried out at rather high doses. Since humans are normally exposed to very low concentrations of chemicals, one has to extrapolate the results obtained at high doses to the low-dose range. Furthermore, to obtain a dose-reponse curve for humans, an extrapolation from the test species to humans also has to be made.

In extrapolating from animals to humans differences in the biological mechanisms of action underlying tumor induction, such as uptake and transport of the chemical in the body, metabolism, and potential detoxification processes, must be taken into account. In practice, interspecies extrapolation boils down to determining equivalent doses. That is to say, if a certain dose causes a specific disease incidence among the test animals, one has to find out what dose will cause the same incidence in humans. Different scientists and agencies advocate different approaches to this problem. The most commonly applied procedure is to assume that the equivalent dose is proportional to the species' body weight or to its body surface (see Chapter 4).

In extrapolating from the high- to the low-dose range some mathematical model of response has to be employed. Many such models have been proposed in the literature, among which the multistage model is probably the most frequently applied (Day and Brown, 1980). Some regulatory agencies use just a simple linear model in order to ensure conservative estimates for the potential health effects in humans (see Chapter 4).

Owing to our limited understanding of the mechanisms causing cancer in animals and humans, there is currently no consensus among scientists about what is the best method for transforming animal results to corresponding human data. As various methods may give vastly different risk estimates, use of animal data to determine dose-response relationships in humans is an uncertain business. For example, in dose extrapolation from mouse to man, use of body weight and body surface gives results differing by a factor of 10 or more. In the high- to low-dose extrapolation uncertainties may be even greater, some models giving risk estimates differing by several orders of magnitude relative to other models (see section 3.1).

Finally, mention should be made of a method proposed by Lewtas (1986) in which short-term mutagenicity tests and epidemiological data are used to obtain cancer risk estimates for humans. The method, called the comparative potency method, is based on the hypothesis that there is constant relative potency between two different carcinogens across different bioassay systems, including humans. Being a hybrid of the indirect and epidemiological methods, the comparative potency method avoids the extrapolation problems discussed above. Some of the assumptions made in the method are questionable, however, and the results must be considered equally uncertain as those obtained using other methods.

3 UNCERTAINTIES IN HEALTH RISK QUANTIFICATION

As stated in the previous section, uncertainties in health risk quantification are large. Generally speaking, the origins of the uncertainty are of three different kinds: (1) the processes involved are inherently stochastic in nature, or at least so complex that it is infeasible to build and apply precise deterministic models; (2) plausible functional relationships connecting important variables or coefficients have been established, but some of the coefficients entering the relationships are not known with certainty and cannot be readily determined; and (3) the physics, chemistry, biology or other scientific aspects of the problem involved are incompletely understood so that reliable mathematical models relating pertinent variables cannot be formulated.

Of these three types of uncertainty only the last two present a real problem in health risk analyses. The second type will be referred to here as "parameter uncertainty" while the third one will be called "model uncertainty." Model and parameter uncertainty may occur in all parts of a health risk assessment, both in the exposure assessment and in the dose-response assessment. In sections 3.1 and 3.2 techniques for dealing with the different types of uncertainty are discussed.

3.1 Model Uncertainty

Model uncertainty, i.e., ignorance about the true functional form of the relationship between important variables occurring in the risk calculation, derives from a lack of basic knowledge about the processes underlying the investigated risk. Morgan (1983) mentions three approaches that can be taken to deal with this kind of uncertainty:

Approach 1:   Make a single-model estimate, choosing as a model one that appears reasonable, and, if possible, try to capture the model uncertainty by introducing uncertainty into the model parameters.

Approach 2:   Identify the range of possible alternative models. Perform a separate analysis for each alternative. Then, if it is possible to attain subjective judgments about the relative likelihood of the different models, combine the results to obtain an overall probabilistic model.

Approach 3:   Perform an order-of-magnitude bounding analysis to determine how a change in model form affects the final risk estimate and the conclusions drawn from that estimate.

As may be inferred from section 2.2, the indirect method of establishing dose-response relationships for humans is profoundly associated with model uncertainty. The implications of this, with regard to the accuracy of risk estimates based on the indirect method, may be illustrated by a study of Campbell et al. (1982). In this study, the cancer risk to humans caused by exposure to perchloroethylene (PCE) was estimated on the basis of results obtained in animal experiments. Uncertainty in the estimates was treated roughly along the lines described in Approach 2 above.

The procedure employed by Campbell and coworkers to determine a dose-response relationship for humans consists of three steps: (1) choice of test animal and animal experiment on which to base the determination, (2) choice of scaling factor for determining equivalent doses in humans, and (3) choice of model for extrapolating the dose-response relationship to the doses that humans are exposed to in their environments. Uncertainty is introduced in all of these steps. To account for the uncertainty, Campbell and coworkers formulate two alternative submodels for each step. Two animal experiments are considered, one with rats and one with mice. Two

Table 2.1   Dose-response Models and Estimated Lifetime Risks from PCE Exposure

Dose-response relationship	Step 1 Test animal	Step 2 Scaling factor	Step 3 Extrapolation	Estimated lifetime risk [1]
1	mouse	body weight	linear	$3 \cdot 10^{-3}$
2	mouse	body weight	quadratic	$1 \cdot 10^{-5}$
3	mouse	body surface	linear	$4 \cdot 10^{-2}$
4	mouse	body surface	quadratic	$2 \cdot 10^{-3}$
5	rat	body weight	linear	$2 \cdot 10^{-4}$
6	rat	body weight	quadratic	$8 \cdot 10^{-7}$
7	rat	body surface	linear	$1 \cdot 10^{-3}$
8	rat	body surface	quadratic	$3 \cdot 10^{-5}$

[1] At an assumed dose of 70 mg PCE per day

principles for scaling doses, body weight and body surface, are applied, and both a linear and a quadratic model are used to extrapolate from high to low doses. Following Approach 2 above, all possible combinations of these submodels, eight in number, are then investigated. Table 2.1 shows the range of models considered and the calculated lifetime risks corresponding to a particular dose of PCE.

As can be seen, the models give widely varying estimates of the cancer risk. At the dose 70 mg PCE per day, which is a dose that workers in the dry cleaning industry are often exposed to, estimates range from $8 \cdot 10^{-7}$, obtained with model 6, to $4 \cdot 10^{-2}$, obtained with model 3. The lowest and highest estimates thus differ by a factor of 50,000.

All three steps involved in the risk estimation contribute to the observed variability. Using body surface as the scaling principle always gives a more conservative risk estimate than does body weight (as long as the test animal is lighter than a human). At doses of about 70 mg PCE per day, the model based on body surface gives risk estimates that are roughly 100 times higher than those obtained using body weight in the case of the mouse and roughly 10 times higher in the case of the rat. At the same dose the mouse data also give higher estimates than those obtained with rats, the difference being of the order 10 to 100. This difference is heavily dependent, however, on the specific results obtained in the experiments employed by Campbell and coworkers, and it is not justified, on the basis of these results, to say anything general about which animal will give the most conservative estimate. The linear extrapolation model generally gives a more conservative estimate for human risk than does the quadratic model. In the present example the difference is of the order 100 at doses of about 70 mg PCE per day.

In the study of Campbell et al., the alternative submodels introduced in establishing an overall dose-response relationship were assigned weights on a subjective basis according to their presumed relevance for predicting response in humans. These weights were used to assign probabilities to the overall risk estimates obtained with each of the eight models. Building on these results, Campbell and coworkers performed a decision-analytic analysis of different options to reduce exposure to PCE.

3.2 Parameter Uncertainty

If, in a risk assessment, little is known about the processes that underlie the health effect considered, model uncertainty is likely to be the major concern. In some cases, however, there may be sufficient scientific evidence for making some sort of quantitative statement about the relationship between the basic process parameters. In such circumstances, uncertainty in the estimated quantities will be the main problem.

Concentrating on the dose-response problem in the first place, quantitative assessments usually take one of two forms: (1) an assumption is made about the functional form of the dose-response relationship, and the parameters entering the function are estimated; or (2) direct estimates are made of the fraction of the population suffering the health effect (i.e., the probability of response) at a few specific dose levels. In the former case uncertainty is contained in the parameters that define the relationship (their exact values not being known); in the latter case uncertainty lies in the estimated probabilities. Good examples of both approaches can be found in the work of Sievering and Roberts (1982).

The major question is, of course, how these uncertainties, and the uncertainties associated with the exposure assessment, will affect the overall health risk estimate, frequently measured in terms of expected number of disease cases in the studied population. To look into this, one has to integrate the relevant exposure and dose-response models and propagate the uncertainties in each component model through to the final health effect variable. Usually, the propagation of uncertainty is done by either of two methods, sensitivity analysis or probabilistic analysis.

3.2.1 Sensitivity Analysis. Of the two methods mentioned above sensitivity analysis is the most simple one to perform and the one that requires the least information. One simply varies the uncertain quantities (model parameters or probability estimates) within reasonable ranges, usually one at a time, to find out how changes in these quantities affect the final health risk estimate. This will produce a range of possible values for the overall risk and will also reveal which parameters are most crucial in determining the size of the risk.

3.2.2 Probabilistic Analysis. This technique is considerably more complex than sensitivity analysis as it involves establishing probability distributions over the range of possible values for the uncertain parameters rather than just specifying intervals for these parameters. The probability distributions should be obtained from scientists possessing expertise in the pertinent disciplines. Using their professional judgments, it may be possible to arrive at complete (subjective) probability curves for the parameters in question. At a less ambitious level, one may assume a specific functional form of the distribution and ask the expert to estimate the parameters that define the distribution.

Once probability distributions have been obtained for all relevant quantities, a probability distribution for the overall health risk can be determined by incorporating the distributions into the integrated health risk model. This may be done by use of analytical or numerical techniques. If the uncertainty of the individual parameters is characterized in terms of discrete distributions, probability trees can conveniently be used to obtain an aggregate distribution (see, for example, the United Nations (1984) report on sulphur pollution). If continuous probability functions are at hand, stochastic simulation is an appropriate technique to use. In a stochastic or Monte Carlo simulation values for the uncertain parameters are drawn randomly according to their respective probability distributions. Having determined a set of parameter values in this way, the corresponding value of the final output variable of the model is computed. This procedure, i.e., drawing values for the uncertain parameters and computing the value of the output variable, is repeated a large number of times, often several thousand times. The calculated values of the output variable are then used to establish a frequency distribution for that variable. The procedure is illustrated schematically in Figure 2.1.

In most real situations the number of uncertain parameters is quite large. In such cases a full probabilistic analysis, along the lines sketched above would be impractical. One may then carry out a preliminary sensitivity analysis to acquire information on what parameters are particularly important and therefore should be subjected to a detailed analysis, and what parameters could be treated less thoroughly when analyzing uncertainty.

To illustrate the use of stochastic simulations in health risk quantification, a specific example of a such an analysis is discussed in the next section.

3.3 Estimation of Excess Angina Attacks Caused by Exposure to Carbon Monoxide

When people breathe air containing carbon monoxide (CO), CO combines with the haemoglobin in the blood to form carboxyhaemoglobin (COHb). With the CO concentrations that are common in streets with heavy traffic, heart patients will experience an increased frequency of angina attacks due to the elevated levels of COHb. An attempt is made here to quantify the additional number of heart attacks occurring per day in Oslo as a result of CO exposure.

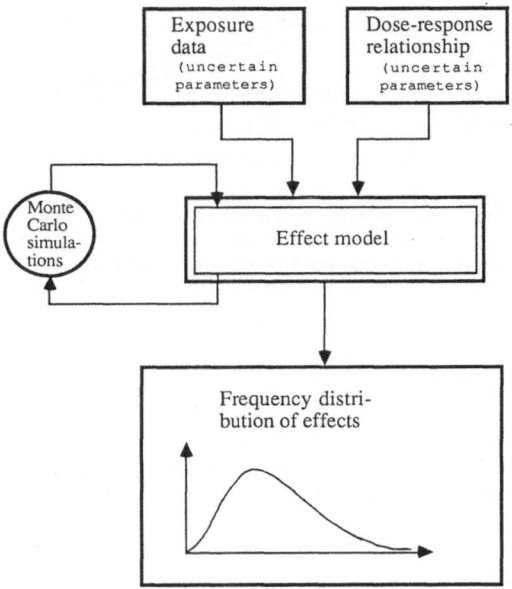

Figure 2.1   Procedure for determining frequency distribution of health effects

The exposure of the population in Oslo to CO has been estimated by Aune (1982). According to his results, about 11,000 persons are exposed to CO concentrations greater than 7 mg/m^3 (8 hours averaging times), while very few people are exposed to concentrations higher than 29 mg/m^3. The exposure estimation also shows that the number of persons exposed decreases with increasing CO concentrations.

The exposure data, in the interval between 7 and 29 mg/m^3, have been fitted to a distribution function of the form

$$f_0(x) = k_0 + k_1 \cdot x + k_2 \cdot x^2 \quad , \tag{2.1}$$

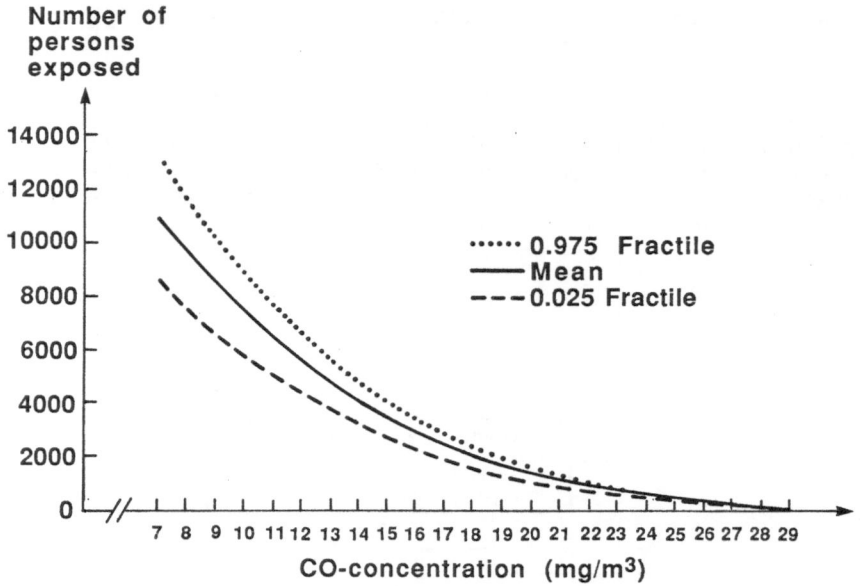

Figure 2.2 Cumulative exposure function for carbon monoxide. For each concentration level the curve gives the number of persons exposed to CO concentrations above that level

where x is the CO concentration in mg/m^3 and $f_0(x)$ is the mean number of persons exposed to concentrations between $\bar{x}$ and x+1 mg/m^3. The coefficients k_1 through k_2 have been estimated to:

$$k_0 = 2275.7$$
$$k_1 = -153.6$$
$$k_2 = 2.7.$$

In this example, uncertainty in the exposure data was estimated subjectively and modelled by a multiplicative error term in equation (2.1). Thus, the expression adopted for the number of persons exposed is:

$$f(x) = f_0(x) \cdot e = (k_0 + k_1 \cdot x + k_2 \cdot x^2) \cdot e , \qquad (2.2)$$

where the error term $\underline{e}$ is normally distributed with mean 1 and standard deviation 0.1. A plot of the cumulative exposure function corresponding to equation (2.2) is shown in Figure 2.2 together with its 95% confidence interval.

Keeney et al. (1984) present a dose-response function relating the probability for a heart patient to have an additional angina attack to the level of COHb present in the blood. By asking several medical experts, they obtained subjective judgments about the likelihood of angina attacks at different COHb levels. These data were then fitted to a logistic function given by

$$P'_A(x') = \exp(a + b\cdot\ln(x') + u)/[1 + \exp(a + b\cdot\ln(x') + u)] , \qquad (2.3)$$

where $P'_A(x')$ is the probability of getting an additional angina attack at the COHb-concentration x', x' being given as per cent of haemoglobin in the circulating blood that is bound to CC, a and b are constants and u is a normally distributed error term with mean 0 and standard deviation 0.5. The dose-response relationship (2.3) is shown graphically in Figure 2.3.

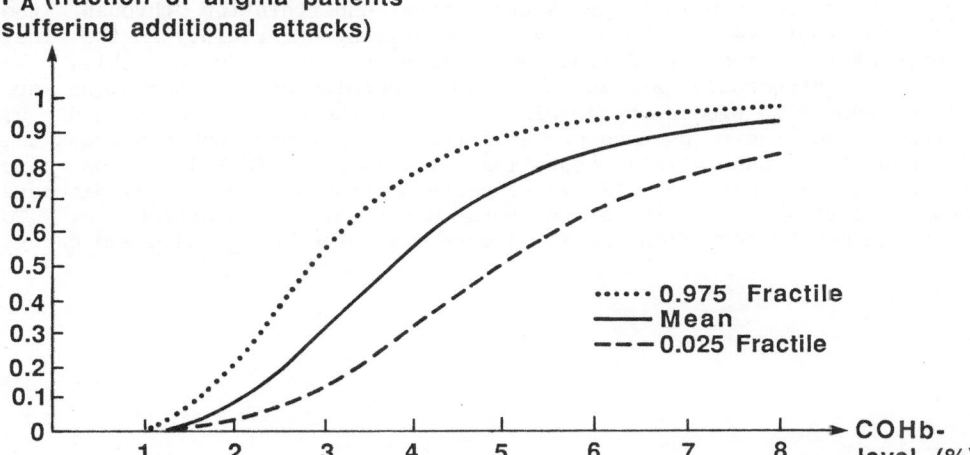

Figure 2.3   Dose-response curve for CO-induced angina attacks in angina patients

To calculate the number of additional angina attacks resulting from CO exposure, we need a model that relates ambient CO concentrations (x) to COHb levels in the blood (x'). In this example, a simple linear relationship,

$$x' = 0.5 + 0.1\cdot x , \qquad (2.4)$$

valid for CO concentrations in the range from 7 to 29 mg/m^3, was developed on the basis of the data given by Aune (1982). No attempt was made to quantify the uncertainty in this relationship.

Combining this relationship with equation (2.3), we can calculate $P_A(x)$, i.e., the probability of getting an angina attack at a specified ambient concentration of CO. The number of additional angina attacks, N, can then be calculated by multiplying the number of heart patients exposed to CO concentrations in a small interval with the corresponding probability $P_A(x)$, and integrating over the range of CO concentrations. This gives

$$N = \alpha \int_{7}^{29} P_A(x) \cdot f(x) dx \, , \qquad\qquad (2.5)$$

where α is the fraction of the exposed population suffering from heart disease. In this analysis we assume, as a rough estimate, the fraction of heart patients to be 3.5%.

The exposure data, as well as the dose-response relationship, contain uncertainties. Our aim is to propagate these uncertainties through the model so that we can quantify the resulting uncertainty in the final estimate of the number of CO-induced angina attacks. To do this, we perform a stochastic simulation. The principles of this technique were outlined above. We proceed by drawing values of the error terms $\underline{e}$ and $\underline{u}$, according to their assumed probability distributions, and construct the functions $f(x)$ and $P'_A(x')$ [equations (2.2) and (2.3)]. Then the integration in equation (2.5) is performed (as a summation of discrete terms) to give a first value of the total number of angina attacks N. To obtain another "observation" of N, we draw new values of $\underline{e}$ and $\underline{u}$ and carry

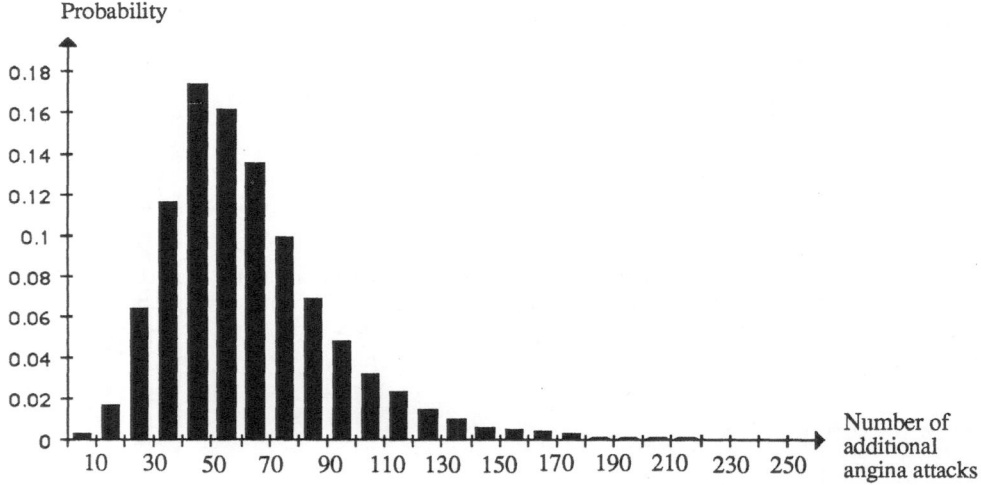

Figure 2.4   Frequency distribution of the number of daily
CO-induced angina attacks in Oslo

out the above computations once more. This process is repeated until we get a reasonably smooth distribution of N. In this example the procedure was replicated 10,000 times. The resulting frequency distribution of the number of angina attacks is shown in Figure 2.4. The mean value of this distribution is 64.2, its standard deviation is 32.2.

4   CONCLUSION

The case of CO discussed in section 3.3 gives an indication of how one would like to proceed in quantifying health effects caused by exposure to chemicals. Unfortunately, the CO case is not very typical of the situation generally encountered in risk assessments, partly because the studied effect, angina attacks, is an acute effect, whereas most health effects of concern, including cancer, develop over a long period of time. In fact, a reasonably accurate characterization of the cancer risk resulting from common ambient exposure levels seems beyond the reach of current possibilities. More research is needed before we are in a position to achieve that goal.

Another factor that complicates chemical risk assessment is the fact that in real life people are exposed not to a single chemical but to a complex mixture of chemicals. Usually, the concentrations of many of the chemicals in these complex mixtures are not known. Furthermore, the toxicity of a particular chemical may be significantly modified (enhanced or reduced) in the presence of other chemicals. For example, the risk of contracting cancer from exposure to asbestos is markedly higher in smokers than in nonsmokers. The possibility of synergistic effects implies that risk estimates for individual chemicals may be of limited relevance in a realistic exposure situation.

However daunting the task may appear, some sort of risk assessment must be undertaken when one is faced with chemicals having the potential of causing adverse health effects in humans or the natural environment. In the absence of more accurate scientific information, it is common practice in many countries to base regulation of chemicals and pollution control on maximum acceptable exposure levels, such as daily intake limits or threshold values for ambient pollutant concentrations. Most of the limit values in use are derived in a fashion that is believed to give a high degree of protection to humans or the environment (see, for example, Chapter 4 of this book).

The use of exposure limits in analyzing chemical risk has several disadvantages. For example, in the case of carcinogens, it is generally believed that there is no absolutely safe concentration level. While it may be very small, a risk of acquiring cancer does exist with any concentration. Furthermore, in the case of health effects for which no-effect levels have been established, a limit value only enables one to say whether or not an effect is likely to occur. Exceeding a threshold, for instance, may allow one to state that premature death or morbidity will very likely occur, but it does not tell us anything about the number of cases. Certainly, in the risk assessment it makes a big difference whether the number is just 1 or say 100. In spite of this, use of threshold values or environmental quality criteria is currently the most widely used approach to regulating chemical pollution. In Chapter 11 of this book, a case study on countermeasures to urban air pollution is described in which application of exposure limits is a central part.

5 REFERENCES

Ames, B.N., 1979, Identifying environmental chemicals causing mutations and cancer, Science, 204:587.

Aune, T., 1982, "Health Effects from Air Pollution in Oslo" (in Norwegian), The Norwegian State Pollution Control Authority, Oslo, Report No. 41.

Campbell, G.L., Cohan, D., and North, D.W., 1982, "The Application of Decision Analysis to Toxic Substances: Proposed Methodology and Two Case Studies," Economics and Technology Division, Office of Toxic Substances, U.S. Environmental Protection Agency, Washington D.C.

Day, N.F., and Brown, C.C., 1980, Multistage models and primary prevention of cancer, J. Nat. Cancer Inst., 64:977.

Dybing, E., 1986, Predictability of human carcinogenicity from animal studies, Regul. Toxicol. Pharmacol., 6:399.

Keeney, R.L., Sarin, R.K., and Winkler, R.L., 1984, Analysis of alternative ambient carbon monoxide standards, Management Science, 30:518.

Lewtas, J., 1986, A quantitative cancer risk assessment methodology using short-term genetic bioassays: the comparative potency method, in: "Risk and Reason: Risk Assessment in Relation to Environmental Mutagens and Carcinogens," P. Oftedal and A. Brøgger, eds., Alan R. Liss, New York.

Morgan, M.G., 1983, "The Role of Decision Analysis in the Implementation of Environmental Policies," OECD, Geneva, ENV/CHEM/CH/83.5, Annex.

Prentice, R.L., and Breslow, N.E., 1978, Retrospective studies and failure time models, Biometrika, 65:153.

Resnick, M.A., 1987, "Evaluation of Short-Term Tests of Genotoxicity," paper presented at the International Conference on Structure-Activity Relationships (SAR) for Toxicological Estimation of Chemicals, Pisa, Italy, May 1987.

Sievering, H., and Roberts, H.A., 1982, "Risk Assessment for Environmental Management: A Case Study of Fuel Switching at Illinois Power Plants," Illinois Department of Energy and National Resources, Springfield, Illinois, Document No. 82/14.

United Nations, 1984, "Air-borne Sulphur Pollution: Effects and Control", Economic Commission for Europe, United Nations, New York, Sales no. E.84.II.E.8.

METHODS FOR ASSESSING THE RISK OF ENVIRONMENTAL CONTAMINATION

C.L. van Deelen

TNO, Netherlands Organization for Applied Scientific
Research, Division of Technology for Society, Department of
Industrial Safety, Apeldoorn, the Netherlands

1   INTRODUCTION

Industrialization has resulted in the increased use of a wide variety
of chemicals. Contamination of the environment may occur at any stage of
the life cycle of a chemical, i.e. during manufacturing, storage, use and
disposal. After release a chemical will initially be transported within
the environmental compartment in which the emission occurs. Depending on
its basic properties a released chemical generally will be transformed to
a certain extent as a result of physical, chemical and/or biological processes. Moreover, in most cases the chemical will enter other environmental compartments, resulting in subsequent transport and transformation.
Regarding the environmental compartments a distinction can be made between
abiotic compartments such as air, (ground)water and soil and biotic compartments, e.g. human beings, flora and fauna.

The pattern of exposure of an organism to a chemical largely depends
on the fate of the chemical in the environment. Examples of potential
exposure routes include inhalation, ingestion via food or drinking water,
dermal contact etc. It has been thoroughly recognized by now that exposure to chemicals may have detrimental effects on human beings, animals
and vegetation. The extent of these effects and their probability of occurrence are referred to as environmental risks.

Decisions on environmental issues, either by the authorities, or
industry, are required in order to maintain environmental risks at an
acceptable level. Such decisions on environmental issues may involve the
reduction or removal of certain emission sources, prohibiting the introduction of new chemicals on the market etc. The instrument of risk
analysis has proved to be a powerful support to decision-making in the
field of industrial and nuclear safety. In view of the above it is beyond
question that risk analysis techniques will be applied to an increasing
degree in order to quantify the impact of the release of chemicals on the
environment and to evaluate the effects of measures that may be taken.
This paper provides a brief survey of the various elements of an environmental risk analysis study. It is pointed out that at present, in many
instances, tools are insufficiently available to enable a detailed quantification of environmental risks to be made. Some of the major problem
areas are highlighted in this context. However, it should be realized

that a detailed environmental risk analysis is not required in all circumstances in order to facilitate decision-making on environmental issues. This is illustrated on the basis of two studies related to soil contamination carried out by the TNO Department of Industrial Safety.

2   ENVIRONMENTAL RISK ANALYSIS

2.1 General

In this paragraph the major elements of an environmental risk analysis are described. Prior to this the features of an environmental risk analysis are compared with those of a safety risk analysis concerning accidental releases of toxic chemicals. In a safety risk analysis the following methodology is usually followed:
*   identification of undesirable events (= release of toxic chemicals);
*   quantification of the consequences of the identified undesirable events (effect and damage analysis);
*   quantification of the probability that the identified undesirable events will occur;
*   presentation and evaluation of the risk an activity represents.

As for the release of toxic chemicals, a safety risk analysis generally has the following characteristics:
*   the release results from a technical and/or operational failure; hence the emission is incidental in character;
*   the consequences of a release are limited spatially: the location of an emission is properly defined and the distance up to which detrimental effects may occur amounts to a few kilometers at the most;
*   the consequences of a release are limited in time, in the sense that generally only acute detrimental effects following a single exposure are quantified;
*   the exposure route is unambiguous: inhalation of a toxic chemical present in the atmosphere;
*   owing to the relatively short exposure time, transport of a chemical from the atmosphere to other environmental compartments, as well as transformation by chemical, physical or biological processes is not taken into account.

Environmental risks are predominantly brought on by continuous emissions. On stating the above it should be realized that there are examples of incidental releases leading to environmental disasters, e.g. Seveso (see also Chapter 8 in the present volume), Chernobyl, and the recent fire at the storage premises of Sandoz in Basel. However, a major distinction between a safety and an environmental risk analysis is that the latter takes both acute and chronic effects into account. From the above it will be clear that an environmental risk analysis is concerned with considerably longer time scales than a safety risk analysis is. Consequently, transformation of a chemical, as well as transport to other environmental compartments can play an important part in an environmental risk analysis.

The geographic dimensions within which the impact of a released chemical on the environment manifests itself may extend in orders of magnitude beyond the immediately affected area in the case of a calamity. An example of an impact at global level is the greenhouse effect due to the worldwide emissions of carbon dioxide, methane, halogenated hydrocarbons and the like. Finally, a safety risk analysis is generally only concerned with damage to human beings and structures, whereas an environmental risk analysis attempts to quantify all adverse effects for the environment as a whole.

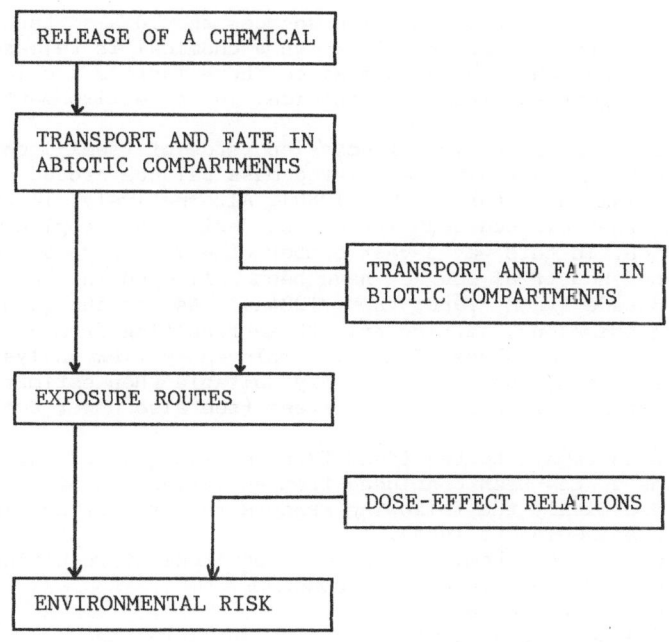

Figure 3.1 Schematic survey of an environmental risk analysis

From the above it will be clear that the quantification of the impact of exposure of a chemical will be considerably more complicated for an environmental than for a safety risk analysis. Figure 3.1 provides a schematic survey of the major elements of an environmental risk analysis. These elements are further discussed below. For a more extensive survey, reference is made to Chapter 1 and Van Deelen and Golbach (1986).

2.2 Release of a Chemical

With regard to the release of chemicals into the environment a distinction can be made between (semi-) continuous and incidental emissions. The latter frequently result from calamities or abnormal (process) conditions and can be considered as highly local emissions (point sources). Additionally, in our society numerous activities produce a (semi-)continuous release of chemicals, e.g. industrial installations, waste disposal sites, agriculture. Some of these emission sources will be highly localized, such as waste disposal sites, others will be diffuse, for example agricultural sources. Some emissions will show a pattern of scattered point sources e.g. domestic waste water, traffic. Irrespective of emission pattern, the identification (where, how?) and quantification (how much?) of emission sources are vital elements in an environmental risk analysis.

An identification of (semi-) continuous sources may be conducted in various ways. In the case of an existing activity, available emission data can be used. When this data are unavailable, identification of emission sources can take place through expert judgement, checklists etc. or through comparison with similar units or activities. The identification of incidental emissions requires systematic analysis of the activity under consideration. For this purpose, techniques can be used that are frequently applied in the field of industrial safety studies, in particular in hazard and operability studies (DGA, 1979).

The identification of emission sources should also take into account the environmental compartment in which a chemical is released as well as its physical and chemical nature since these factors are important parameters for transport and fate of a chemical in the environment.

The quantification of emission rates of (semi-) continuous sources may be obtained from a calculated mass balance around an operation, or from emission factors (VROM, 1980). Alternatively, in the case of a local source the released quantity may be derived by sampling and analysing an emission. In this way a vast number of emission data from both industrial and non-industrial sources have been collected in the Netherlands over the last decade (VROM, 1980; VROM, 1984a). As for the quantification of incidental emissions, particularly those resulting from a calamity, it is recommended to make use of models employed in risk analysis studies. Risk analysis techniques are also very suitable when estimating the probablity or frequency of incidental releases (see also under section 3.1).

With regard to the identification and quantification of emission sources it has been concluded from an evaluation of existing techniques that especially the following aspects require further investigation (van Deelen and Golbach, 1986):
* release of chemicals into the environment resulting from use and disposal of products by consumers;
* emissions to soil;
* frequency and quantity of incidental releases.

2.3 Transport and Fate of Chemicals in the Environment

After release a chemical will be distributed into the environment. The distance over which a contamination may spread, the velocity of transport and the concentration pattern in the environment depend upon a large number of factors, such as:
* quantity of released material
* emission pattern (e.g. in which environmental compartment)
* chemical, physical and biological properties of released chemical(s)
* local conditions (e.g. type of soil, level of groundwater, water renewal rate of a lake or estuary etc.).

To date many models have been developed for the purpose of predicting transport and fate of contaminants in the environment. The complexity of these models varies from simple equilibrium or transport models to complicated, dynamic, multi-compartment models. However, the application of most of the models is restricted to specific conditions, and in many cases the quantification by models of transport and fate of a contaminant in the environment is not yet possible. Moreover, the user's requirement with regard to accuracy and operability of a model is a conflicting one in most cases. Excellent reviews of existing transport and fate models for air, water and soil are given in Swann & Eschenroeder (1983) and Ricci (1985). Examples of application of such models to environmental problems are given in the Chapters 7 and 10 of the present volume.

When a modelling approach can not be followed, it is still possible to estimate the concentration of contaminants in the environment on the basis of available monitoring data. However, estimating concentrations from monitoring data has some disadvantages:
* single monitoring data provide information on the concentration at a specific moment and a specific location; in order to obtain an insight into trends (for example fluctuations in time) an extensive monitoring program is required;

*   unsuitable for new chemicals;
*   less suitable for non-homogeneous environmental compartments such as soil;
*   time-consuming and expensive.

With the exception of models based on the fugacity concept, hardly any models are available to quantify transport and fate of a contaminant over several environmental compartments. The fugacity models are based on the principle that a chemical, when released in a specific environmental compartment, has a tendency to be distributed over the other environmental compartments until an equilibrium has been reached. Based on this principle Mackay and his co-workers have developed different models with an increasing level of complexity (Mackay, 1979; Mackay et al., 1983). The fugacity models, however, represent a limited field of application in that they only quantify the distribution of a contaminant over the basic environmental compartments air, water, soil, sediment and biota rather than the concentration gradients within each of these compartments. Consequently, these models are most suitable in environmental risk analysis studies at a global or regional level. As an initial step the distribution of a contaminant over the environmental compartments can be checked roughly by means of relatively simple fugacity models. After this initial check, specific models can be applied to describe in greater detail transport and fate of a contaminant in a particular environmental compartment.

2.4 Exposure Routes

An organism can be exposed in various ways to a contaminant in the environment. An exposure route is made up of the medium in which a contaminant is transported and the manner in which exposure takes place e.g.
*   inhalation of gases or particles;
*   ingestion of drinking-water or food;
*   dermal contact.

The transport media include, among other things air, food, drinking-water and surface water (recreational facilities). In an environmental risk analysis an integrated evaluation has to be made of all potential exposure routes with respect to their relative importance and subsequently the major ones have to be analysed in greater detail.

Models for the quantification of an exposure dose in particular have been developed in the field of risk analysis for nuclear activities. The models require a thorough evaluation with regard to their applicability in environmental risk analysis. Similar to risk analysis techniques for industrial safety, the development of models, taking into account the frequency distribution of the level of exposure for a population, requires further attention. These models are especially required for analyses of site-specific environmental risks. Such exposure models should include behaviour aspects of (parts of) the population, such as consumer's behaviour, occupation etc.

2.5 Dose-effect Relations

An essential step in an environmental risk analysis constitutes "translating" into effect the estimated dose to which an organism is exposed. Ideally, such a translation should be done on the basis of a quantitative dose-effect function relating the fraction of a population suffering from certain effects (e.g. mortality) to the level of exposure. However, quantitative dose-effect functions as defined above are only available for a few chemicals. Moreover, the vast majority of currently available dose-effect functions relate to the short-term effects of acute exposure. Data on the long-term effects of acute exposure and the effects of chronic exposure is still very scarce at present.

As for the chronic exposure to non-carcinogenic chemicals, it is generally accepted that they will only cause adverse effects when a specific threshold value is exceeded. Toxicity research on non-carcinogenic chemicals has therefore been directed at determining the "no-effect level" of such chemicals. Where these compounds are concerned, hardly any dose-effect relations are available thus making it impossible to quantify the probability of occurrence of detrimental effects with regard to a certain population. An environmental risk analysis of non-carcinogenic contaminants will therefore only provide insight into the probability that the estimated level of exposure will exceed a pre-fixed value, e.g. a no-effect level.

The existence of a threshold value for carcinogenic and mutagenic chemicals is still a matter of dispute among scientists. The pros and cons for a threshold value for these chemicals are given in a review article by Crump (1985a). Authorities in the Netherlands as well as in the United States (Environmental Protection Agency) have adopted the non-threshold philosophy in assessing the risk of carcinogens and mutagens.

Data on the carcinogenicity and mutagenicity of a chemical may be derived from laboratory studies on human or animal cells, from toxicological laboratory studies on whole animals and from epidemiological and clinical studies of human populations. When dealing with risk analysis studies for human beings it is generally preferred to make use of epidemiologic data derived from human populations when available. In following this approach a common problem is that exposure of the human population is never fully characterized (Morris, 1987). Moreover, epidemiologic data may have been influenced by other, unknown carcinogenic factors.

The application of animal studies results in a human risk analysis study introduces a number of important uncertainties (Morris, 1987; Crump, 1985b):
* high to low dose extrapolation
* extrapolation from short-term to long-term exposure
* animal to man extrapolation, taking into account body weight, which breathing volume etc.
* differences in metabolism
* extrapolation from one route of exposure to another

Several models exist for quantifying the probability of cancer at low doses on the basis of short-term experiments at high doses (Crump, 1985b; Fiksel and Scow, 1983). The most commonly applied model is the so-called "one-hit" model, which assumes a linear proportional relation between the level of exposure to a carcinogen and the cancer incidence. It should be realized that models such as the "one-hit" model are a simplification of the actual mechanism of a carcinogen. This is for example illustrated by the fact that the individual sensitivity for a specific carcinogen may vary several orders of magnitude, depending on genetic factors, occupation, consumer behaviour etc.

At present knowledge about effects of toxic compounds on non-human organisms at population and eco-system level is still in its very early stages. In addition to the above there is a serious lack of knowledge on the effects of simultaneous exposure to different chemicals in general. The same goes for individual sensitivity towards (combined) exposure to toxic chemicals.

2.6 Evaluation of Environmental Risk Analysis Tools

From the above outline it will become clear that, with today's knowledge, a detailed environmental risk analysis remains out of reach for most conditions. In this context a detailed environmental risk analysis represents the full quantification of the environmental impact of a chemical release.

On the other hand, it is unrealistic to postpone decisions on environmental issues till adequate models have been developed. The next section illustrates how risk analysis techniques can be applied in order to support decision-making.

3 EXAMPLES OF A SIMPLIFIED APPROACH

3.1 Risk Analysis of Underground Storage Tanks

In the Netherlands a large number of underground storage tanks for industrial use are situated at filling stations. On the basis of available data, it may be stated that leakages from underground storage tanks are major contributors to cases of soil contamination (VROM, 1984b). For that reason the Ministry of Housing, Physical Planning and the Environment commissioned TNO to carry out research into the risk of soil contamination in consequence of leakage of petrol and diesel oil from underground tanks at filling stations. Two storage systems have been compared in the study, i.e. a single-walled tank in accordance with Dutch directives and a double-walled tank in accordance with German directives. The results of the study have been reported more extensively in Van Deelen (1986), and in full detail in Pietersen and Van Deelen (1985).

The risk assessment concerns the storage tank (including its coating and cathodic protection), as well as the filling line, filling point and filling procedure for transshipment from the tank car to the tank. The leak detection system has in addition been considered for the double-walled tank. The risk assessment methodology may be represented schematically as follows:

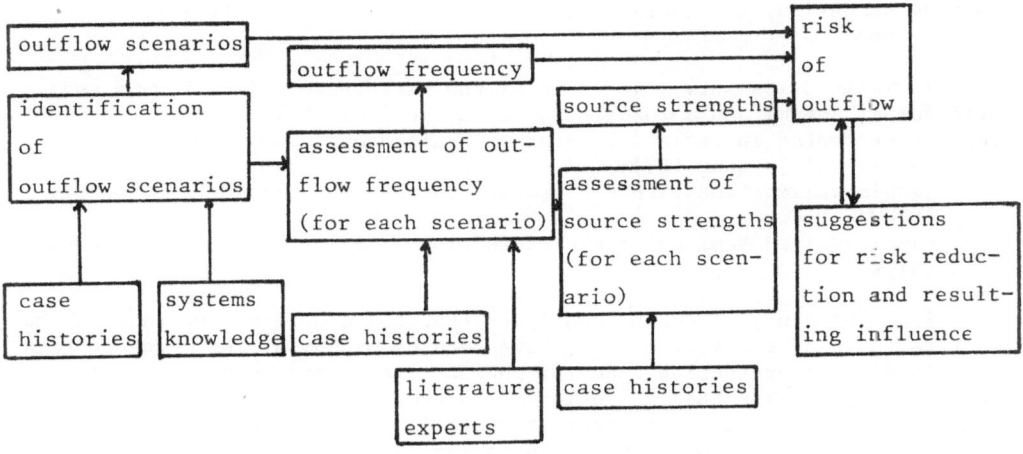

The identification of outflow scenarios has been carried out on the basis of case histories and a systematic analysis of the storage/filling system. The following outflow scenarios have been identified in the process:
(1) external corrosion
(2) internal corrosion
(3) outflow during transshipment
(4) leakage from mechanical damage
(5) leakage from design/construction faults
(6) outflow when tanks are being put out of operation

The leakage frequency of oil products has been assessed for each of the identified outflow scenarios. In assessing the failure frequencies, case histories - for the greater part - and data from literature - in a few cases - were made use of. Where possible, the assessment of failure frequencies has taken into consideration (site-) specific parameters such as aggressiveness of the soil, presence of overfill protection, type of stored product etc.

In the cases where only insufficient or incomplete information was available, assumptions and marginal conditions have been formulated. The obtained failure frequencies were subsequently submitted for checking to experts in the field of underground storage of mineral oil products. The velocity at which petrol or diesel oil spills from an underground tank in cases of leakage is dependent on a number of factors: amount and size of the holes, level of groundwater etc. The total released quantity (source strength) however is, apart from these factors, even more determined by the fact that outflow often continues until a certain nuisance level has been reached (e.g. pollution of drinking water, stench nuisance etc.). On the basis of an analysis of case histories of soil contamination the average source strength has been estimated for each outflow scenario.

The combination of outflow frequency and resulting amount of oil product that is released constitutes a measure for the risk of soil contamination resulting from an outflow scenario. Many minor leakages lead in the end to a degree of soil contamination comparable to one major spill, because oil products tend to accumulate in the soil. Therefore it was decided to represent the risk as a product of the frequency of occurrence and the amount of outflow (risk factor). The risk factors provide an insight into the degree to which the various outflow scenarios contribute to the overall risk. Moreover, by totalling the risk factors of individual outflow scenarios, the overall risk of various storage systems to the soil can be compared.

In Table 3.1 a survey is given of the risk factors for storage and transfer of petrol and a resistance of the soil of 75-100 ohm m. The results presented in Table 3.1 refer to 3 different situations:
* double-walled tank (DT);
* single-walled tank within a protection zone for public water supply (STWW);
* single-walled tank outside a protection zone for public water supply (STOW).

From the study it can be concluded that the double-walled tank is the most reliable of the systems that have been examined. A single-walled tank positioned in a protection zone for public water supply constitutes a somewhat higher risk than a double-walled one. This is caused by the absence of a protection system against overfill and increased leakage frequency from internal and external corrosion.

Table 3.1 Risk Factors in Relation to Soil Contamination from
Underground Petrol Tanks at Filling Stations

	DT	STWW	STOW
Leakage from external corrosion	0.2	<0.1	10.0
Leakage from internal corrosion	<0.1	1.0	2.0
Outflow during transhipment	5.4	7.4	7.4
Leakage from mechanical damage	0.3	0.3	0.3
Faulty installation/construction	0.3	0.3	0.3
Outflow during excavation	<0.1	<0.1	<0.1
Total	6.2	9.0	20.0

Of the examined systems, a single-walled tank positioned outside a
protection zone for public water supply constitutes the greatest risk.
The above applies especially to tanks which have not been fitted with
cathodic protection. In these cases there is a relatively high leakage
frequency from external corrosion caused by the fact that after the tank
has been installed not a single check takes place anymore on the condition
of its coating and the presence of a galvanic contact or electrical influence from outside (stray current/interference).

On the basis of the results of the risk analysis the study also
provides a number of risk reducing measures, the major ones of which are:
*   more stringent rules with regard to the installation for cathodic
     protection systems;
*   periodical checks on the condition of the coating of all tanks, i.e.
     also those which have not been fitted with cathodic protection;
*   increased frequency of inspection of the cathodic protection system
     for tanks positioned in soils with a high degree of aggressiveness;
*   increased frequency of internal inspection, allowing for differen-
     tiation between petrol and diesel oil tanks;
*   installation of a protection system against overfill.

From the above it will be clear that no attempt has been made in the
study to quantify the effects of a release of oil products, for example in
terms of damage to the soil. However, in view of the fact that the storage
systems under consideration contain a similar type of product, the simplified approach in this study provides a proper scientific basis for the
development of guidelines for underground storage tanks in general. In
case guidelines have to be drawn up for specific conditions, the local
circumstances will have to be taken into consideration as well and under
those circumstances the released quantity may have to be translated into
environmental impact one way or the other. This aspect is described in the
next section.

3.2 Assessing the Risk of Soil Contamination from Industrial Activities

The method that has been applied in assessing the risk of soil contamination from underground storage tanks has been further developed into
a procedure to quantify the risk of soil contamination from industrial
activities in general (van Deelen, 1987). The objective of the procedure
is to pinpoint, from a wide variety of chemicals, activities, operations
and process apparatus, the potential problem areas related to soil contamination from an industrial activity, and to quantify the associated

risks. The procedure takes into account both "regular" emissions to soil and releases due to calamities (e.g. failure of equipment, procedural errors). The developed procedure basically consists of 5 major steps:

(1) Inventory of chemicals, activities, process apparatus etc. in an industrial activity and identification of potential problem areas regarding soil contamination;

(2) Analysis of identified problem areas and definition of scenarios that may actually lead to cases of soil contamination;

(3) Quantification of the frequency of occurrence and the released amount of chemicals for each of the defined scenarios;

(4) Quantification of a risk factor in relation to the location of an industrial activity;

(5) Presentation and evaluation of the risk of soil contamination.

Assessment of the risk of soil contamination with the developed procedure goes one step further than the risk analysis of underground storage tanks does, in the sense that the location of an industrial activity is taken into account. For that purpose risk factors have been derived that constitute a relative measure for the volume of contaminated soil in the case of a leakage. In the procedure the following parameters are taken into account when quantifying the risk factor of a location:

* permeability of the soil, which is related to the natural character of the soil;

* maximum distance over which a contaminant may be transported; this distance is made dependent on the geohydrological conditions and the distance between the source of release and the nearest open water course.

From the above it will be clear that the quantification of a risk factor with regard to surroundings involves a number of significant simplifications. For example, the risk factors have been made independent of type and quantity of released material; processes such as adsorption/desorption, dissolving/depositing and decomposition/conversion are therefore unaccounted for. However, it should be realized that transport and fate of a chemical depends on its chemical, physical and biological properties in relation to the nature of the soil. It will be clear that within the scope of a general applicable procedure it is impracticable to distinguish between all possible combinations of chemicals and types of soil. Moreover, the risk factors have not been developed for the purpose of quantifying the concentration profile of a contaminant in the soil but are only required to provide a relative measure of the potential volume of soil that may be contaminated. For that reason the procedure does not take into account properties of chemicals such as toxicity, tendency to accumulate in biomass etc.

Application of the procedure to an industrial activity provides in a quantitative way, an insight into the extent to what the various parts of an installation or activity contribute to the risk of soil contamination. On the basis of a quantification of risks it is possible to draw up measures to reduce the risk of soil contamination. Finally, the procedure enables a comparison to be made of the risk of soil contamination from different industrial activities.

With regard to assessing the risk of soil contamination, the procedure has been applied quite successfully in the case of a Dutch company - Chemco Europe NV, Soest - producing photochemicals and photosensitive emulsions. This company is located in the near vicinity of a groundwater pump station. The pumped up groundwater is directly used for public water supply, and any degree of contamination of the groundwater will render it unsuitable for this purpose. For that reason the regional authorities have imposed severe regulations on the company in order to minimize the risk of soil contamination. Implementation of all the provisions imposed in the guidelines would involve a considerable amount of investment. In view of this, the company was allowed to introduce the provisions in phases, on the basis of the results of a risk analysis study. With the procedure developed by the TNO Department of Industrial Safety, the potential sources of soil contamination at the company site have been identified and the associated risks quantified. As for the potential sources of leakage, which constitute a relatively high level of risk, recommendations were made regarding measures in order to reduce the risk of soil contamination. In this context it is interesting to note that these risk-reducing provisions differed considerably from those adopted in the regulations.

4   CONCLUSIONS

Risk analysis techniques have proved to be a powerful support to decision-making in the field of industrial and nuclear safety. It has been demonstrated that the quantification of the impact of long-term exposure to chemicals is of a considerable complexity, and that a detailed environmental risk analysis is not quite feasible in a large number of circumstances. This paper gives a brief survey of the areas requiring further development in this respect. Meanwhile decisions on environmental issues frequently have to be made on the basis of urgency and cannot await the development of models to carry out detailed environmental risk analysis studies. In those circumstances decisions have to be made on the basis of a limited number of data. This paper gives some examples of simplified risk analysis methods related to soil contamination. It is demonstrated that the results of such simplified methods can serve to support decision-making.

5   REFERENCES

Crump, K.S., 1985a, Mechanisms Leading to Dose-Response Models, in: "Principles of Health Risk Assessment", P.F. Ricci, ed., Prentice-Hall/Englewood Cliffs, New Jersy.

Crump, K.S., 1985b, Interspecies Comparison for Carcinogenic Potency to Humans; in: "Principles of Health Risk Assessment", P.F. Ricci, ed., Prentice-Hall/Englewood Cliffs, New Jersey.

Deelen, C.L. van, and Golbach, G., 1986, "Programming Note on Environmental Risks" (in Dutch), TNO report 86-070.

Deelen, C.L. van, 1986, Assessing the risk of soil contamination in the case of industrial activities, in: "Contaminated Soil", J.W. Assink, and W.J. van den Brink, eds., Martinus Nijhoff Publishers, Dordrecht.

Deelen C.L. van, 1987, "A procedure for assessing the risk of soil contamination by industrial activities" (in Dutch); to be issued by the Government Publishing Office, the Hague.

Directorate-General of Labour (DGA), 1979, "Hazard and Operability Study. Why? When? How?", Report of the Directorate-General of Labour, 1st edition.

Fiksel, J.R., and Scow, K.M., 1983, Human Exposure and Health Risk Assessment Using Outputs of Environmental Fate Models, in: "Fate of Chemicals in the Environment", R.L. Swann, and A. Eschenroeder, eds., ACS Symposium Series 225.

Mackay, D., 1979, Finding Fugacity Feasible, Environmental Science and Technology, 13: 1218.

Mackay, D., Paterson, S. and Joy, M., 1983, Application of Fugacity
Models to the Estimation of Chemical Distribution and Persistence
in the Environment, <u>in</u>: "Fate of Chemicals in the Environment",
R.L. Swann and A. Eschenroeder, eds., ACS Symposium Series 225.

Morris, S.C., 1987, Dose-Response Curves for Acute and Chronic Exposure
and Response to Chemical and Physical Agents, Paper Prepared for
the Workshop for the Presentation of Guidelines for the Joint
IAEA/UNEP/ WHO Project on "Assessing and Managing Health and Environmental Risks from Energy and Other Complex Industrial Systems", held at Brookhaven National Laboratory, June 29-July 2,
1987.

Pietersen, C.M., and van Deelen, C.L., 1985, "Comparative risk analysis of
underground storage systems at (car)filling stations" (in Dutch).
"Bodembeschermingsreeks" no. 45. Government Publishing Office, the
Hague.

Ricci, P.F.(ed.), 1985, "Principles of Health Risk Assessment", Prentice-
Hall/Englewood Cliffs, New Jersey.

Swann, R.L., and Eschenroeder A. (eds.), 1983, "Fate of Chemicals in the
Environment", ACS Symposium Series 225.

VROM, 1980, "Handbook of Emission Factors, Part 1 - Non-industrial
sources", Government Publishing Office, the Hague.

VROM, 1984a, "Handbook of Emission Factors, Part 2 - Industrial Sources",
Government Publishing Office, the Hague.

VROM, 1984b, Provisional indicative soil program for 1984-1988 (in
Dutch).

METHODS USED IN THE UNITED STATES FOR THE ASSESSMENT AND MANAGEMENT

OF HEALTH RISK DUE TO CHEMICALS

J. W. Falco

Office of Environmental Processes and Effects Research
United States Environmental Protection Agency
Washington, D.C.

R. V. Moraski

Office of Health and Environmental Assessment
United States Environmental Protection Agency
Washington, D.C.

1   INTRODUCTION

Assessing the risk from either deliberate or accidental environmental chemical releases is a key factor in developing a strategy for the control of environmental pollution or the protection of human health.   A risk assessment also may be used to estimate the degree of risk reduction that could result from the consideration of control scenarios that may be implemented in the regulatory process.   Over the years legislation and regulations have been enacted in the United States to control the release of potentially hazardous chemicals into the environment and to institute risk reduction strategies.   The result has been a diversity of approaches and technical quality in risk assessments which have complicated the management of environmental risks.   Consequently, several efforts have been undertaken in an effort to gain an understanding of the risk assessment/risk management process and to reduce the multiplicity of approaches used by the various federal agencies.

This paper reviews the recent major U.S. reports on the issues of risk assessment and risk management and the response of federal agencies to the recommendations made in those reports.

2   NATIONAL RESEARCH COUNCIL, 1983

A major report was published by the National Research Council (NRC) of the United States National Academy of Sciences (NAS) in 1983 (NRC, 1983).   In its report, "Risk Assessment in the Federal Government: Managing the Process," two elements of regulatory actions--risk assessment and risk management--are described.   A summary of this description is as follows.   During the risk assessment process, the potential adverse health effects to individuals or populations are identified.   The prediction of the extent of exposure and the determination of the number and characteristics of persons exposed at various intensities and durations are assessed. A characterization of the uncertainties involved in all of the estimates is an important part of risk assessments.

The risk management process evaluates alternative regulatory actions and chooses among them the most appropriate response to a potential acute or chronic health hazard. In making the selection, the risk manager considers political, social, and economic inputs along with the risk assessment.

The steps in developing a risk assessment include:

1. Hazard Identification. This step attempts to determine whether exposure to a chemical agent causes an increase in the incidence of a health effect, and whether the evidence is strong enough to characterize the agent as a potential health hazard. Since the available data which show effects of chemicals on humans are usually limited, positive tests in laboratory animals may be used as evidence that the chemical agent may pose a risk to exposed humans.

Four general classes of information may be used in attempting to identify the hazard: epidemiologic data, animal bioassay data, data on in vitro effects, and comparisons of molecular structure and activities. Epidemiologic studies, if well conducted, should show whether there is a positive relationship between the chemical agent and the adverse health effect. Even though the evidence gathered in epidemiologic studies would be the most convincing evidence about the risk to humans, the difficulty of obtaining statistically meaningful data at the time of release of the agent to the environment poses limitations on the use of such data and requires the use of less direct evidence that a health risk exists.

Animal bioassay data are the most commonly available data used in hazard identification. The fundamental inferences that result from animal experiments are applicable to humans if considered valid and accepted. In some cases, interpretation of such data has been difficult, but these tests generally have proven to be reliable indicators of carcinogenic properties, as well as other properties such as reproductive effects.

Other supportive evidence comes from comparing a chemical's physical-chemical properties and structure and chemical reactivity with that of known carcinogens. If the comparisons indicate effects, further investigation may be needed before identifying the chemical as a health hazard.

The NAS compiled a list of components in carcinogenic risk assessments. As many as 25 components in hazard identification were listed. The question of the degree of scientific uncertainty in each component and the effect of choosing an inference option are most critical in the hazard identification process. The overall weight of evidence of carcinogenicity must be considered by judging the overall quality of the data presented to support the conclusion reached.

2. Dose-Response Assessment. This step characterizes the relationship between the dose of the agent either administered or received and the incidence of the health effect. This assessment takes into account such variables as intensity of exposure, activity patterns of the persons exposed, and other factors which may affect the relationship. The process usually requires extrapolation from high to low dose and extrapolation from animals to humans.

As in the hazard identification step, the absence of human data is common. Tests performed on rats, mice, or other animals must be evaluated in a dose-response assessment. Dose-response assessments frequently require that response curves be generated from limited animal test data. In addition, obvious size and metabolic differences between humans and laboratory animals require that the results be adjusted to reflect these differences.

Many aspects of dose-response assessments identified by the NAS introduce uncertainty. Three aspects that introduce major uncertainties are high- to low-dose extrapolation, interspecies conversion of dose, and the decision on whether to combine tumor types in determining tumor incidences. Different extrapolation methods could produce results that vary by several orders of magnitude.

3. Exposure Assessment. The exposure assessment is the process for describing, measuring, and estimating the intensity, frequency, and duration of human exposures from existing chemicals in the environment or from the release of new chemicals into the environment, the routes of exposure, and the size and nature of the human population exposed. While exposure assessments based on actual data are preferred, such exposure data may be incomplete, and the assessor may have to rely on models for estimating exposure. With new chemicals, reliance on data for similar chemicals may be necessary.

Current approaches to exposure assessment appear to be medium- or route-specific. The generation of an exposure assessment may be complicated when multimedia effects need to be considered. However, the assessor has several options available to estimate the exposure in a specific medium, which will yield more or less conservative estimates. The same process is then applied to the other media of concern, and the exposures are added together.

4. Risk Characterization. The risk characterization describes the nature and the magnitude of human risk under various conditions of human exposure. It is accomplished by combining the exposure and dose-response assessments, including the analysis of the uncertainties underlying the assessments.

In all of these steps, data and assumptions contain varying degrees of uncertainty. In the dose-response assessment, statistical and biological uncertainties need to be characterized. The accuracy of the exposure assessment is dependent on the quality and quantity of the data and the type and complexity of the mathematical models used. For each exposure pathway assessed, a discussion of the strengths and weaknesses of the methodologies used to generate the exposure estimates should be presented. A thorough discussion of these attendant uncertainties is an important part of the overall risk assessment process.

2.1 Inference Guidelines

The NAS report recommended the use of inference guidelines for risk assessment. These guidelines are intended to be principles followed by risk assessors in interpreting and reaching judgments based on scientific data. One advantage of using guidelines, according to the NAS report, is that use of guidelines would help keep the risk assessment separate from other parts of the regulatory process. That is, it is hoped that prior regulatory conclusions would not influence the risk assessment process. Comprehensive and detailed guidelines would insulate the components of a risk assessment and provide a sharp distinction between risk assessment and risk management.

A second advantage to guideline use is that it ensures the application of quality control. If risk assessments are based on judgments formulated by experts in diverse fields such as epidemiology, chemistry, toxicology, etc., then knowledge gaps can be bridged and judgments made based on state-of-the-art thinking in each field.

A third argument for the use of guidelines is consistency. Consistency ensures fairness and rationality. Guidelines should help avoid inadvertent omissions. Consistency will enhance comparability of results with respect to assumptions and characterization of the uncertainty in the estimates made.

A fourth advantage is predictability. By knowing how the government will evaluate health effects data, industry can assess its own activities and testing programs. Conflicting policies and unanticipated changes in judgments would be minimized.

By providing a focus for continual debate and examination of the options generally used, revisions in the inference guidelines options will ensure that the guidelines' utility in assessing risk will increase. In addition, detailed guidelines that set forth scientific and policy bases of risk assessment could foster improved public understanding of government regulatory actions. The use of inference guidelines would mean that risk assessments would not need to reevaluate the same issues with every assessment. Repetitious discussions could be reduced because policy determination and current state-of-the-art scientific understanding would not be repeatedly addressed.

Some of the potential disadvantages cited in the NAS report of developing and using guidelines include the following:

Guidelines would encourage a rote approach to risk assessments. Scientific judgments would give way to risk assessment by formula. However, risk assessments do not lend themselves to nongeneric interpretation, and the need remains for the risk assessor to make judgments.

The use of guidelines would make the mixing of science and policy unavoidable, as in the use of the most conservative dose-response curve, and this mixing would go unrecognized by the general public. Carefully designed guidelines, however, should distinguish between scientific knowledge and policy and reduce the opportunity for unrecognized mixing.

Considerable time and money would be required to develop guidelines and would detract from an agency's ability to regulate. However, if guidelines define elements of risk assessment policy that are subject to change and scientific elements that are not subject to change based on policy decisions, then a derived benefit is stability of the regulatory process.

The guidelines may freeze science by hindering the timely incorporation of new scientific thinking. Guidelines that are flexible and adopted as procedures would allow incorporation of new scientific evidence based on scientific justification.

2.2 Guideline Development: A History

The following is a brief history of guideline development for carcinogenic risk in the United States.

1. International Agency for Research on Cancer (IARC), 1971. Guidelines used for evaluation of potential carcinogens are set in general terms and are related to six components of hazard identification. These guidelines featured the presentation of criteria that classified evidence of suspected carcinogens as sufficient or limited.

2.   Food and Drug Administration (FDA), 1973, 1977, 1979.   These guidelines specified a 10^{-6} lifetime cancer risk as a quantitative criterion of significance.   These were the first guidelines to define significant risk.   Later in this paper the current risk assessment philosophy of the Department of Health and Human Services (DHHS), of which the FDA is a part, is discussed.

3.   United States Environmental Protection Agency (EPA), 1976.   Early EPA guidelines were referred to as cancer principles.   These early guidelines incorporated a two-step process.   The first step is a determination of whether a substance constitutes a cancer risk.   This is followed by the determination of the regulatory action, if any, that should be taken to reduce risk.   A discussion of the the EPA's current guidelines follows later in this paper.

4.   Occupational Safety and Health Administration (OSHA), 1977. These guidelines addressed hazard identification but did not address exposure assessment and spurned the use of dose-response assessment.

5.   United States Consumer Product Safety Commission (CPSC), 1978. The cancer guidelines developed by the CPSC dealt with hazard identification and listed 10 components of this process.   Some attention is given to exposure assessment and dose-response assessment.

6.   Interagency Regulatory Liaison Group (IRLG), 1979.   This group, representing four agencies (CPSC, OSHA, FDA, and EPA), produced the most comprehensive guidelines at the time and represented interagency consensus on scientific aspects of carcinogen risk assessment.   Those guidelines addressed most components of hazard identification and dose-response assessment, and contained limited general discussion of exposure assessment and risk characterization.   The White House Office of Science and Technology Policy (OSTP) succeeded this group in 1981.   In 1985, the OSTP published a document on the scientific basis of risk assessment.   The document is discussed later in this paper.

2.3 NAS Conclusions and Recommendations

The NAS concluded that guidelines are necessary, feasible, and desired.   Guidelines should accommodate change and be viewed as aids, not formulas.   Guidelines for health effects other than cancer would be based on less extensive data bases.

Among the overall recommendations made in the NAS report were the following:

- Risk assessment should be separated from risk management;

- Risk assessment should be conducted before deciding whether a substance is a health hazard;

- Risk assessments should be reviewed by an independent advisory panel; and

- Guidelines should be developed that are comprehensive, detailed, and flexible.

3   COMMITTEE ON SCIENCE AND TECHNOLOGY, 1983

The Committee on Science and Technology of the U.S. House of Representatives published a report prepared by the Congressional Research

Service (CRS) entitled "A Review of Risk Assessment Methodologies" (CRS, 1983). The report examines the technical elements of risk analysis, along with the uses and limitations of each major method used in risk assessment. The report discusses such areas as laboratory research, including animal and cellular research; epidemiologic studies; systems models, including probabilistic analysis; biological, environmental, and ecological transport processes; and comparative analytical techniques, used by decision-makers as approaches to risk management.

The primary conclusion reached by this committee is that the risk assessment process is subject to many uncertainties and errors. An evaluation of these uncertainties is essential. It is suggested that worst-case comparisons, rather than mid-range figures, be used when comparing alternative risks in order to be certain that errors lie on the side of public safety. Further, identifying research that may reduce uncertainties could result from an understanding of the limitations and capabilities of the risk assessment. To further reduce uncertainties, a weight-of-evidence approach is suggested as a means of reinforcing the findings from other research involving the same risk agent.

4 OFFICE OF SCIENCE AND TECHNOLOGY POLICY, 1985

The Office of Science and Technology Policy (OSTP) document defines principles, based on scientific data, to serve as guidelines for assessing carcinogens (OSTP, 1985). Risk is composed of two aspects: hazard and exposure. Hazard refers to the toxicity of the substance, which is deduced from a wide array of data, including epidemiologic evaluation of long-term animal studies, short-term studies, and structure-activity relationships. Exposure is defined as the amount of substance with which humans come in contact.

The steps of a risk assessment presented by the OSTP are similar to those found in the NAS document:

1. Evaluation of the evidence that cancer will occur (hazard identification).

2. Estimates and distributions of human exposure (exposure assessment).

3. Estimates of dose-response relationships at low doses (dose-response assessment).

4. Combining of exposure assessment with dose-response assessment to obtain estimates of risk (risk characterization).

Since assumptions and approximations are made because of limitations in existing data bases, treatment of the uncertainties is required in each step. The report defines an unreasonable risk as a risk that outweighs the benefits associated with the use of a chemical substance.

The document is divided into two parts. The first part presents the principles relative to the role of chemicals in carcinogenesis. The second part contains detailed information supporting the principles presented in Part 1. The issues discussed in the second part include the mechanisms of cancer induction, short-term and long-term tests for potential carcinogens, epidemiologic methods, exposure assessment methods, and the steps in the risk assessment process. Among the principles discussed are the following:

Carcinogenesis is characterized as a multistage phenomenon. The presence or absence of a threshold for one step of the carcinogenic process does not determine whether a threshold is present or absent for the whole process. Short-term tests of cancer induction serve as screens for potential carcinogens, and those adequately validated should be selected to characterize the response of the chemical. In the absence of adequate human data, it is reasonable to regard chemicals for which there is sufficient evidence for animal carcinogenicity as if they presented a risk of cancer in humans.

Epidemiologic methods provide the only means for directly assessing the carcinogenic risk of environmental agents to humans. However, well-designed, well-conducted, and well-evaluated studies can provide a basis for causal inferences that can be used in public health decisions. A negative epidemiologic study does not prove the absence of an association between exposure to the agent and human cancer.

Exposure assessments should be based on monitoring data, physical and chemical models, and consideration of all exposure routes through all media. The assessment should describe the strengths, limitations, and uncertainties of the data used, and models should state assumptions used to generate estimates. A range of exposure estimates is preferred to a single estimate.

In deriving the risk assessment, all relevant data should be considered, using sound biological and statistical procedures. No single mathematical procedure is recommended for low-dose extrapolation. The risk assessment is no more accurate than the exposure assessment it uses. Qualitative as well as quantitative estimates of uncertainty should be made when considering strengths and weaknesses of the risk assessment.

5   UNITED STATES DEPARTMENT OF HEALTH AND HUMAN SERVICES, 1985

This report was written by the Committee to Coordinate Environmental and Related Programs for the United States Department of Health and Human Services (DHHS) as a complement to the NAS and OSTP documents (U.S. DHHS, 1985). The Public Health Service Act gives the Secretary of the DHHS the authority to protect and promote the health of people in the United States. Many agencies within the DHHS have specific responsibilities in the protection of health and in the risk assessment/management of various toxic substances. The work of these agencies is described in the following paragraphs.

With respect to toxic substances, the Alcohol, Drug Abuse, and Mental Health Administration is engaged in risk assessment of the use of psychotherapeutic drugs. The National Institutes of Health supports and conducts research that includes the toxicology of environmental contaminants as it relates to human disease.

Within the National Institutes of Health, the National Institute of Environmental Health Sciences explores the health effects of chemical substances by being primarily involved in hazard identification and hazard characterization. The research focuses on developing fundamental knowledge of the mechanisms of chemical toxicity and on improving risk assessments by developing biomarkers in laboratory and epidemiologic studies.

The National Cancer Institute identifies high-risk groups and individuals, and occupational and environmental causes of cancer. Some of the research of this group centers on the role of tumor initiators and promoters and cocarcinogens in cancer risk.

The Centers for Disease Control conduct research in the health effects of exposure to toxic agents, and identify human exposure to these agents. The Centers are involved in hazard identification and analysis of the risk assessment process.

The Agency for Toxic Substances and Disease Registry (ATSDR) has responsibilities mandated under the Superfund Amendments and Reauthorization Act (SARA) and the Resource Conservation and Recovery Act (RCRA) to determine the extent of a health risk to a population as a result of a release of a hazardous substance into the environment. In addition, the ATSDR is mandated to establish registries of diseases and people exposed to toxic substances in the environment, to establish a data bank on the toxicity of chemicals, and to study the relationship of illness to exposure to toxic substances.

The Food and Drug Administration is mandated by the Federal Food, Drug, and Cosmetic Act to conduct and participate in many risk assessment/management activities to meet its responsibility to ensure the safety of food, drugs, cosmetics, biological products, radiation-emitting electronic products, and medical devices.

The Office of the Assistant Secretary of Health administers the largest hazard identification program in government, provides statistical data for risk assessments, and performs risk assessments as needed. The National Toxicology Program (NTP), which is a part of this office, develops, validates and applies test methodologies needed to produce better chemical toxicological profiles.

The four steps of the risk assessment process as defined by the NAS and OSTP reports are followed by the DHHS. Hazard identification relies primarily on clinical and epidemiologic studies, as well as on animal toxicological experiments and tests that shed light on a chemical's mechanisms of action. Exposure assessment is based on monitoring data, modeling, study of microcosms, or a combination of some or all of these. Monitoring studies can focus on the home, the workplace, or the ambient environment. Exposure may vary with the population, may occur concurrently with other chemicals, and may be subject to differing levels of bioavailability. Bioavailability of a given chemical may be difficult to determine. Absorbed dose may not correlate with external exposure levels.

The techniques employed most frequently for the dose-response assessment are the safety-factor approach, margin-of-safety procedure, and mathematical modeling. The safety-factor approach is used primarily for noncarcinogens to determine acceptable or allowable intake levels. The margin of safety approach uses the estimated no-observed-effect level in the most sensitive species tested and divides that number by the estimated human exposure to produce the margin of safety. This provides a qualitative indicator of the proximity of the exposure level to that which may cause an adverse effect. Mathematical modeling is used to depict dose-response relationships to determine estimates of and upper-bounds on the low-dose risks which are not known.

Other assumptions used in the risk assessment by the DHHS include the following: no threshold for carcinogenesis exists; risks from multiple exposures and multiple sources of exposure to the same chemical are additive; in the absence of data, 100% absorption in all species is assumed regardless of the route of exposure; and in the absence of pharmacokinetic data, the effective dose is assumed to be proportional to the administered dose.

In the risk characterization step the overall determination of the potential health risk is made by evaluating all data on hazard, exposure, dose-response, and species extrapolation. This is accompanied by an evaluation of the uncertainties in each step of the assessment process.

The DHHS procedure for risk management depends on statutory, scientific, organizational, and political constraints imposed on the available options. In addition, if the estimates of risk are found acceptable, then priorities for managing the substances are developed based on legal, political, scientific, and social factors. Keeping risk management separated from risk assessment ensures the utility and independence of the scientific component and allows the public to properly weigh any regulatory decisions.

6   UNITED STATES ENVIRONMENTAL PROTECTION AGENCY, 1986*

At the United States Environmental Protection Agency (EPA), risk assessment has come to play an increasingly important role in the regulatory process. The distinction between risk assessment and other parts of the regulatory decision process was clearly defined by the National Academy of Sciences (NAS) in 1983 and can include legal, economic, political, and social factors as a part of the management of risks determined by way of the risk assessment process (National Research Council, 1983).

The EPA has taken a number of steps to help achieve quality and consistency in its risk assessments, including the development of guidelines (guidelines were developed for carcinogenicity in 1976 and 1980, systemic toxicants and mutagenicity in 1980, and exposure assessment in 1983). In January 1984, the EPA began work on six new or revised guidelines: carcinogenicity, mutagenicity, reproductive toxicity (subdivided into individual guidelines for developmental toxicity and male and female reproductive toxicity), systemic toxicants (i.e., target organ toxicants), chemical mixtures, and exposure assessment. Five of these risk assessment guidelines were published in the Federal Register in 1986 (U.S. EPA, 1986a-e).

The EPA's guidelines set forth internal Agency procedures that:

- promote consistency across EPA risk assessments by developing common approaches to risk assessment;

- promote the quality and accuracy of the science underlying the EPA risk assessments by use of a consensus approach (discussed below) where appropriate; and

- clarify the EPA's approach to risk assessment by informing the public and the regulated community about the process by which the EPA will evaluate scientific information.

The guidelines are not regulations; they are intentionally general to allow for the need to use appropriate scientific methods and judgments. However, they impact the regulatory process by:

- making the EPA's risk assessments more accurate and of higher technical quality;

- familiarizing risk assessors throughout the country with the EPA's approach; and

---

*The material in this section has been abstracted in part from Preuss and Ehrlich, 1987.

- making it possible for scientists to plan their experiments to collect the information EPA scientists would like to have available when conducting a risk assessment.

These guidelines are intended to be evolving documents. They will be updated, as the science base relating to risk assessment leads to new understanding of the effects of toxic substances, or to a reduction of the uncertainty inherent in the risk assessment process. To the extent possible, the EPA's guidelines follow the NAS's recommendations and will be discussed on that basis.

6.1 Hazard Identification

The hazard identification component of a risk assessment consists of a review of relevant biological and chemical information bearing on whether or not an agent may pose a specific hazard. Sometimes, there is enough information available for the qualitative evidence to be combined into a formal weight-of-evidence determination.

For example, in the Guidelines for Carcinogen Risk Assessment (U.S. EPA, 1986a), the following information is evaluated to the extent that it is available:

- physical/chemical properties and routes and patterns of exposure;

- structure/activity relationships;

- metabolic and pharmacokinetic data;

- the influence of other toxicologic effects;

- short-term tests;

- long-term animal studies; and

- human studies.

Once these data are reviewed, the animal and human data are each divided into groups by degree of evidence:

- sufficient evidence of carcinogenicity;

- limited evidence of carcinogenicity;

- inadequate evidence;

- no data available; and

- no evidence of carcinogenicity.

The animal and human evidence are combined into a weight-of-evidence classification scheme similar to one developed by the International Agency for Research on Cancer (1982). This scheme gives more weight to human evidence when it is available. The scheme includes the following groups:

- Group A - human carcinogen.

- Group B - probable human carcinogen.

- Group C - possible human carcinogen.

- Group D - not classifiable as to human carcinogenicity.

- Group E - evidence of noncarcinogenicity towards humans.

In the case of mutagenicity risk assessment (U.S. EPA, 1986c), the goal is to assess the likelihood that a particular chemical agent induces heritable changes in DNA, and the likelihood that the chemical will inter- act with human germ cells.

Evidence that an agent induces heritable mutations in human beings could be derived from epidemiologic data indicating a strong association between chemical exposure and heritable effects. It is difficult to ob- tain such data because any particular mutation is a rare event, and only a small fraction of the estimated thousands of human genes and conditions are currently useful as markers in estimating mutation rates.

Therefore, in the absence of human epidemiologic data, it is appro- priate to rely on data from experimental animal systems as long as the limitations of using surrogate and model systems are clearly stated. The universality of DNA and the interest in the possible causal relationship between mutagenesis and cancer induction are partly responsible for the development of a large number of both _in vitro_ and _in vivo_ mutational tests that may be used to evaluate the potential mutagenic activity of specific agents. The practical implication is that the data available for any set of chemicals are extremely variable, thus precluding a precise scheme for classifying chemicals as potential human germ-cell mutagens. What has evolved is a rank-ordered scheme of categories of evidence bear- ing on potential human germ-cell mutagenicity. The highest category is reserved for human epidemiologic data, even recognizing that no such data are currently available. There are five categories in descending order based on the premise that greater weight is placed on tests conducted in germ cells than in somatic cells, on tests performed _in vivo_ rather than _in vitro_, in eukaryotes rather than prokaryotes, and in mammalian species rather than submammalian species. Additionally, there is a category for defining a nonmutagen and one for insufficient information for a qualitative decision to be made.

The specific statements of the eight categories are:

1. Positive data derived from human germ-cell mutagenicity studies, when available, will constitute the highest level of evidence for human mutagenicity.

2. Valid positive results from studies on heritable mutational events (of any kind) in mammalian germ cells.

3. Valid positive results from mammalian germ-cell chromosomal aberration studies that do not include an intergeneration test.

4. Sufficient evidence for a chemical's interaction with mammalian germ cells, together with valid positive mutagenicity test re- sults from two assay systems, at least one of which is mammalian (_in vitro_ or _in vivo_). The positive results may both be for gene mutations or both for chromosomal aberrations; if one is for gene mutations and the other for chromosomal aberrations, both must be from mammalian systems.

5. Suggestive evidence for a chemical's interaction with mammalian germ cells, together with valid positive mutagenicity evidence from two assay systems, as described under 4 above. Alterna- tively, positive mutagenicity evidence of less strength than

defined under 4 above, when combined with sufficient evidence for a chemical's interaction with mammalian germ cells.

6. Positive mutagenicity test results of less strength than defined under 4 above, combined with suggestive evidence for a chemical's interaction with mammalian germ cells.

7. Although definitive proof of nonmutagenicity is not possible, a chemical could be classified operationally as a nonmutagen for human germ cells, if it gives valid negative test results for all end points of cancer.

8. Inadequate evidence bearing on either mutagenicity or chemical interaction with mammalian germ cells.

In the the EPA's guidelines, developmental toxicity includes adverse effects on the developing organism that may result from exposure prior to conception (in either parent), during prenatal development, or postnatally to the time of sexual maturation (U.S. EPA, 1986d). The major manifestations of developmental effects include death of the developing organism, malformation, altered growth, and functional deficiency. (The term teratogenicity refers primarily to malformations and is a subclass of developmental toxicity). Short-term and _in_ _vitro_ tests, which are frequently used for assessing risks from suspect carcinogens and mutagens, are not appropriate approaches for assessing developmental toxicity because the developing organism is such a complex system. Instead, bioassays and human epidemiologic data are the primary sources of information used. The primary biological assays involve treatment of animals during organogenesis and evaluation of the offspring at term. These types of evaluations may also be done as part of a multigeneration study.

The kinds of evaluations that are made in the EPA's hazard identification/weight-of-evidence determination include:

● quality of the data;

● resolving power of the studies; that is, consideration of the significance of the studies as a function of the number of animals or subjects;

● relevance of route and timing of exposure;

● appropriateness of dose selection; and

● evaluation of the information for a series of end points which may include:

-- In the developing animal:

deaths
structural abnormalities
growth alterations
functional deficiencies in the developing organism

-- In the maternal animal:

fertility
weight and weight gain
clinical signs of toxicity
specific target organ pathology and histopathology.

In the case of chemical mixtures (U.S. EPA, 1986e), the EPA conducts its hazard identification by considering the weights-of-evidence for the component chemicals of the mixture. Occasionally, and especially for complex mixtures, the evidence for health hazard comes directly from occupational studies on the mixture itself. Information on the mixture itself, however, must be carefully reviewed for evidence of masking. For example, when one of the component chemicals is a suspect carcinogen, but the data show marked toxicity in major organs (e.g., liver, kidney) and no indication of cancer, there is the possibility that other toxic effects may mask the evidence of carcinogenicity. The hazard identification then would suggest no cancer risk at any dose, when in fact, at doses below the threshold for systemic toxicity, there could be a significant risk of cancer.

For mixtures, the exposure information must be considered to determine the chance that chemical interactions in the environment could produce new chemicals, over time or during transport, with different types of health hazards. This is discussed more fully in the next section.

6.2 Dose-Response Assessment

Classically, there are two general approaches to dose-response assessment. In the case of nonthreshold effects (e.g., carcinogens), there is an attempt to develop low-dose extrapolation models that assess the likelihood of risk at the levels to which the population will be exposed. The largest number of such dose-response extrapolations have been performed in the field of carcinogen risk assessment; therefore, the cancer guidelines give the most detailed guidance on dose-response assessment (U.S. EPA, 1986a). These include guidance on the kinds of evidence that should be used in the dose-response evaluation:

- if available, estimates based on human epidemiologic data are preferred over estimates based on animal data;

- in the absence of appropriate human studies, data from animal species that respond most like humans should be used;

- the biologically acceptable data set from long-term animal studies showing the greatest sensitivity should generally be given the greatest emphasis;

- data from the exposure route of concern are preferred to data from other exposure routes; if data from other exposure routes are used in making route-to-route extrapolations, this process must be carefully described;

- when there are multiple tumor sites or multiple tumor types, each showing significantly elevated tumor incidence, the total estimate of carcinogenic risk is done by pooling, i.e., counting the number of animals having one or more of the significant tumors; and

- benign tumors should generally be combined with malignant tumors for risk estimates.

Another major consideration is the choice of the particular mathematical model used for low-dose extrapolation. Different extrapolation models may fit the observed data reasonably well but may lead to large differences in the projected risk at low doses. The EPA will review each assessment as to the evidence on cancer mechanisms and other biological or statistical evidence that indicates the biological suitability of a particular extrapolation model. In the absence of adequate information to the contrary, the linearized multistage procedure will be used because it is recognized as leading to a plausible upper limit to the risk that is consistent with some proposed mechanisms of carcinogenesis.

Additional issues are species and route extrapolation of the doses. Currently, the EPA adjusts animal doses by the ratio of animal-to-human surface areas. The evidence in support of this approach is not strong, and research is in progress to improve the method. Route extrapolation is used when the only available data are for a route different from the route of concern.

In the case of mutagenicity risk assessment (U.S. EPA, 1986c), dose-response assessments can presently be performed only with data on germinal mutations induced in intact mammals. The morphological specific locus and biochemical specific locus assays can provide data on the frequencies of recessive mutations, and data on heritable chromosome damage can be obtained from the heritable translocation test. As in carcinogen risk assessment, the EPA will strive to use the most appropriate extrapolation models for risk analysis and will be guided by available data and mechanistic considerations in this selection. However, it is anticipated that for tests involving germ cells of whole mammals, few dose points will be available to define dose-response functions, and a linear extrapolation will therefore be used. The EPA has recognized that pioneering work in the field of molecular dosimetry may ultimately lead to useful extrapolation models.

The other major approach to dose-response assessment is concerned with effects which the EPA has referred to as systemic toxicants or noncarcinogenic health effects (see below). Although this particular area is not now covered by guidelines (they are still being developed), it is appropriate here to discuss the general approach. The EPA usually calculates a reference dose (RfD), that is, the dose below which a significant risk of adverse effects is not expected. The Rfd is related to the concept of the acceptable daily intake (ADI), but strives to remove the elements of risk management from the process. At present, the EPA is uncertain at what point above the RfD there will be a significant adverse health effect. The dose-response evaluation is done in the following way. The literature is examined to determine both the critical toxic effect (that is, the adverse effect that first appears in the dose scale as the dose is increased) and the highest dose at which the effect does not occur (often called the highest no-observed-adverse-effect level or NOAEL). This NOAEL is divided by an uncertainty factor which generally ranges from 10 to 1,000; the uncertainty factor is composed of a series of factors, each representing a specific area of uncertainty inherent in the data available.

The RfD calculation is a generic calculation for most toxicants considered to have thresholds. In addition, much work is being carried out in an attempt to develop more quantitative approaches for dose-response assessment for reproductive and developmental toxicants both within and outside of the EPA.

The dose-response procedures described in the chemical mixtures guidelines are slightly different (U.S. EPA, 1986e). In this case, guidance is provided for combining several different types of information on the mixture of concern as well as on the mixture's components. If dose-response data are unavailable on the specific mixture, it is appropriate to infer information from sufficiently similar mixtures. When neither type of information is available, it is suggested that what is called dose or response addition be used, with appropriate modifications, if interactions between components such as synergism can be quantified.

For most threshold pollutants that are toxicologically similar, strict dose addition is used. This involves dividing each estimated intake level by its RfD, and summing each of these quotients to calculate a hazard index. When the hazard index is much less than one, no risk is expected from the mixture. When the hazard index is much greater than one, a significant risk is expected. When the hazard index is near one, each case needs to be considered individually.

For carcinogens, and for dissimilar toxicants that have dose-response data, response addition is used so that, at typical environmental levels, the excess risks for each of the component chemicals are summed to reach an overall risk estimate. Again, interactions need to be considered. Performing such a calculation when the individual risk estimates are plausible upper bounds results in the uncertainty introduced being on the side of increased protection of public health. The EPA intends to investigate this and other problems involving mixtures that contain carcinogens over the next few years.

6.3 Exposure Assessment

As can be deduced from the titles of the various guidelines, four of the five relate to health effects; in those cases, which have been discussed previously, discussions of hazard identification and dose-response assessment are appropriate. In contrast, one guideline discusses exposure assessment.

The Guidelines for Estimating Exposures (U.S. EPA, 1986b) provides a procedural framework on how best to estimate the degree of human contact with a chemical. The major areas to be evaluated when estimating exposures are:

- source assessment - a characterization of the sources of contamination;

- pathways and fate analysis - a description of how a contaminant may be transported from the source to the potentially exposed population;

- estimation of environmental concentration - an estimate using monitoring data and/or modeling of contamination levels away from one source where the potentially exposed population is located;

- population analysis - a description of the size, location, and habits of potentially exposed human and environmental receptors; and

- integrated exposure analysis - the calculation of exposure levels and an evaluation of uncertainty.

Generally, exposure estimates may be presented by expressing the magnitude and duration of an individual event of exposure or by expressing potential lifetime exposure. For example, data on acute or subacute effects, such as developmental effects, would use the magnitude of exposure per event or several events over a short period of time. On the other hand, assessments of carcinogenic risk often consider the daily average exposure calculated over a lifetime. The nature of the toxic effect being evaluated in the risk assessment will determine the appropriate length of exposure presented.

For most risk assessments involving chronic exposure, exposure (mg/kg/day) is calculated as a dose averaged over the body weight (kg) and lifetime (days):

$$\text{Average Daily} \atop \text{Lifetime Exposure} = \frac{\text{Total Dose}}{\text{Body Weight X Lifetime}} \qquad (4.1)$$

The total dose (mg) can be expanded as follows:

$$\text{Total} \atop \text{Dose} = \text{Contaminant} \atop \text{Concentration} \quad X \quad \text{Contact} \atop \text{Rate} \quad X \quad \text{Exposure} \atop \text{Duration} \quad X \quad \text{Absorption} \atop \text{Fraction} \qquad (4.2)$$

The four parameters in equation (4.2) are defined as follows:

- Contaminant concentration represents the concentration of the contaminant in the medium (air, food, soil, etc.) contacting the body. Typical units are mass/volume or mass/mass;

- Contact rate is the rate at which the medium contacts the body (through inhalation, ingestion, or dermal contact); typical units are mass/time or volume/time or for dermal contact, volume/surface area/time;

- Exposure duration is the length of time for contact with the contaminant; and

- Absorption fraction is the effective portion of total contaminant contacting and entering the body. Entering the body means that the contaminant crosses one of the three exchange membranes: alveolar membrane, gastrointestinal tract, or skin.

The six factors given in equations (4.1) and (4.2) must be known (or assumed) in order to estimate exposure. Research is in progress as to how to estimate each of these factors for humans as well as test animals.

6.4 Risk Characterization

In the EPA's guidelines, the risk characterization step is a summing discussion in which information is put together in a useful way. This means that the risk characterization contains not only a risk estimate for a specific exposure, but also a cogent summary of the biological information, the assumptions and their limitations, and a discussion of uncertainties in the risk assessment, both qualitative and quantitative.

In the case of the cancer, mutagenicity, and chemical mixtures guidelines (U.S. EPA, 1986a,c,e), the risk characterization specifically consists of a listing or discussion of the dose-response extrapolation information as well as the associated weight-of-evidence determination from the scale or table contained in the guidelines. For mixtures (U.S. EPA, 1986e), the weight of evidence covers three areas: health effects information, toxic interactions, and exposure estimates.

In the case of the exposure assessment guidelines, a specific mathematical technique has been developed for assessing uncertainty (U.S. EPA, 1986b). In this case, the probability distributions estimated for the uncertainty around each of the components in the calculation are tabulated, and the probability distribution of the results of the exposure assessment can then be calculated or estimated.

6.5 Summary

Risk assessment at the EPA has evolved into a general analytic and decision-making tool used by many people in many programs across the Agency. Furthermore, many of the EPA's statutes are now predicated on a risk reduction basis, which requires more and better health risk and exposure analyses. Since the possibility of overlapping and conflicting analyses exists, a larger role is necessary for review of risk assessments in general, for oversight of the process, and for development of more detailed guidelines. The structures for assuring this quality and technical consistency are evolving within the Agency.

Therefore, as the sophistication of risk assessment increases, as more risk assessments are performed, and as there is more need for the assurance of quality and consistency, the EPA will develop guidelines for more end points, add more detail to existing guidelines, strengthen the management procedures for resolving the scientific disputes, publicize those resolutions within the Agency, and maintain the appropriate degree of oversight. The result of the process will be the development of risk assessments with less overall uncertainty, and ultimately, better protection of public health.

7   EXAMPLE OF A RECENT EPA RISK ASSESSMENT*

Recently the EPA completed an intensive review and evaluation of the latest data that have a bearing on the risk assessment of dichloromethane (DCM; methylene chloride). These data, and the most current methods for their analysis, were reviewed in an effort to improve the quantitative risk estimates for DCM. The following paragraphs summarize this effort. This risk assessment represents the EPA's analysis of the weight of evidence regarding DCM's carcinogenic potential for humans; it includes revised cancer risk estimates which take into account the new information submitted to date on pharmacokinetics, mechanism of action, and epidemiology, and it discusses methods for incorporating pharmacokinetic information into the cancer risk assessment. This risk assessment draws on the body of work developed by the Health/Risk Assessment Committee (HRAC), an interagency committee that was established to evaluate the health effects caused by DCM as well as other halogenated solvents.

Dichloromethane is a multipurpose solvent with application in paint stripping, metal cleaning, foam blowing, electronics, and chemical processing. It is also a component in certain aerosol propellant mixtures. Because of its many applications, a large number of people may be exposed to DCM in the workplace, through use of consumer products, or from emissions to ambient air.

Animal bioassays published in the 1980s raised concern that DCM could cause cancer in humans. Several bioassays reported elevated tumor incidences in mice and rats exposed to this chlorinated solvent. In response to these bioassay data and concerns about other potential chronic effects, the EPA's Office of Air and Radiation (OAR) requested that the Office of Health and Environmental Assessment (OHEA) prepare a Health Assessment Document (HAD) on DCM. The HAD, which reviewed data relevant to acute and chronic effects, was published in February 1985 (U.S. EPA, 1985a).

New bioassays published in 1985 by the National Toxicology Program (NTP) intensified concern for DCM's potential to cause cancer. The EPA's Office of Toxic Substances (OTS) began a regulatory review to determine whether the risks posed by any or all uses of DCM were sufficient to warrant priority consideration for regulatory action under section 4(f) of the Toxic Substances Control Act (TSCA). The OAR, already involved in investigating whether ambient sources of DCM should be regulated under the National Emissions Standards for Hazardous Air Pollutants (section 112)

---

*The material in this section has been abstracted in part from U.S. EPA, 1987.

provision of the Clean Air Act (CAA), requested that OHEA prepare an addendum to the HAD to evaluate the new data from the NTP.

In May 1985, OTS published (U.S. EPA, 1985b) a finding under TSCA section 4(f) that DCM may present a significant risk to humans of serious and widespread harm from cancer. The new NTP bioassays provided sufficient evidence to upgrade DCM from a possible to a probable human carcinogen and led to a positive 4(f) finding. Following the 4(f) action, OTS and OAR combined efforts, and a single coordinated assessment was published by OHEA in 1985 as an Addendum to the HAD for DCM (U.S. EPA, 1985c). In the Addendum, the EPA classified DCM as a probable human carcinogen (Group B2) as defined by the EPA Guidelines for Carcinogen Risk Assessment (U.S. EPA, 1986a). The assessment led to publication of an EPA Advance Notice of Proposed Rulemaking in October 1985 (U.S. EPA, 1985d).

The EPA was not alone in responding to the NTP's new bioassay data. Following release of the NTP data, several other federal regulatory agencies began to investigate the health effects posed by DCM. The Consumer Product Safety Commission (CPSC) released a briefing package in June 1985 announcing the staff's intent to pursue voluntary labeling actions. Subsequently, CPSC began a rulemaking procedure to determine whether or not DCM should be designated a hazardous substance, and published a Federal Register notice (U.S. CPSC, 1986) requesting comment on the proposed rule. In November 1985, the Food and Drug Administration (FDA, 1985) proposed a ban on the use of DCM in hairsprays. The Occupational Safety and Health Administration (OSHA) began an investigation into the risks associated with occupational exposures to DCM and published an Advance Notice of Proposed Rulemaking in November 1986 (OSHA, 1986).

The multiple assessments called for a coordinated effort on the part of the various agencies to ensure consistency in regulatory decision-making. To that end, the EPA proposed developing a common integrated assessment covering not only DCM but also other chlorinated solvents that could serve as substitute chemicals as well. The agencies developed an integrated strategy to assess the major occupational, ambient, and consumer sources of exposure and risk from DCM, tetrachloroethylene, trichloroethylene, carbon tetrachloride, methyl chloroform, and CFC-113.

The integrated strategy led to the establishment of a committee composed of representatives from the EPA, CPSC, OSHA, and FDA to assess data relevant to the health risks that might be associated with exposure to the chlorinated solvents. The committee, called the Health/Risk Assessment Committee (HRAC) of the Integrated Chlorinated Solvent Project, was charged with reviewing and updating, if necessary, the existing hazard/risk assessments on the solvents. Separate committees were set up to resolve issues concerning ambient air and occupational and consumer exposure, and to address economic issues through the development of a risk/benefit analysis. The work of these committees is not yet completed.

In response to the NTP bioassay and the federal regulatory agencies' investigations, industry and other external groups developed new data on DCM, focusing primarily on species differences in metabolism and the mechanism of tumor initiation. The HRAC has made it a top priority to evaluate these new data. The HRAC's evaluations of these new data serve as the basis for the EPA's updated risk assessment of DCM (HRAC, 1987).

Data from the 1985 NTP inhalation cancer bioassay demonstrate that DCM is oncogenic in two species of laboratory animals, rats and mice. In rats, tumors were benign fibroadenomas of the mammary gland. The literature (Russo et al., 1982) suggests that adenomas of this type are histogenically different from malignant adenocarcinomas and do not have a high potential for progressing to malignant tumors. As the rat tumors were not of a type with known malignant potential, the relevance of these tumors to human health is unclear. The response in rats cannot be entirely discounted but is considered to carry less weight than the response in mice.

In mice, the response was unequivocally carcinogenic. Administration of DCM via inhalation at doses of 2,000 or 4,000 ppm caused a statistically significant increase in malignant tumors in two organs, the liver and lung, in B6C3F1 mice. Although the tumors were of a type that have occurred at a high and/or variable background frequency in the strain of mouse tested, there is no question of their statistical significance in the NTP study. The tumor increases in mice were dramatic: at 4,000 ppm, 40/50 male mice and 41/48 female mice developed lung tumors, while 33/49 male mice and 40/48 female mice developed liver tumors. Moreover, the tumors occurred in a pattern which meets the criteria for determining whether commonly occurring tumors provide sufficient evidence of animal carcinogenicity; an excess of malignant tumors was observed in both sexes of mice, and the proportion of malignancies to benign tumors, as well as the time to first appearance of tumor, were dose-related. The NTP data thus provide sufficient evidence of animal carcinogenicity.

The EPA Guidelines for Carcinogen Risk Assessment (U.S. EPA, 1986a) and the Office of Science and Technology Policy (OSTP, 1985), generally accepted by the scientific community, recognize that chemical agents for which there is sufficient evidence of carcinogenicity in animals should be regarded as probable carcinogens for humans. The OSTP policy states that a "finding of carcinogenicity in rodents is proof that the chemical is carcinogenic in a mammalian species. Such a finding must be taken as strong evidence that the chemical can be a carcinogen in man, unless there is compelling. . .evidence to the contrary."

The varied responses in chronic animal studies of DCM--a clear carcinogenic response in mice exposed at high doses, a negative response in hamsters, a primarily benign tumor response in rats, and the lack of a statistically significant increase in malignant tumors in mice exposed at low doses--are grounds for investigating whether there are biological differences between species that could lead to species (and/or dose-related) differences in risk.

The EPA has been concerned with evaluating whether the new data on DCM regarding species differences in metabolism and the carcinogenic mechanism of action lead to the conclusion that DCM does not present a cancer risk to humans. The EPA believes that, for such a statement to be made, the data would have to indicate with reasonable certainty that:

- the mechanism of carcinogenic action in mice is not expected to occur in humans; or

- the carcinogenic biochemical pathway is inactive in humans; or

- the epidemiologic data are sufficient to clearly indicate that DCM does not cause cancer in humans.

After a critical analysis of the evidence, the EPA has concluded that the data fall short of meeting any of these criteria. The carcinogenic mechanism of action of DCM remains unidentified. There is little evidence supporting either an epigenetic or a genotoxic mechanism. Results from a large number of genotoxicity studies are mixed. Unequivocal positive results were obtained chiefly in prokaryotic systems; positive results were fewer as the complexity of the test organism increased and test systems progressed from _in vitro_ to _in vivo_. The insensitivity of the available mammalian _in vivo_ studies and the lack of supporting evidence for an alternative mechanism allow the suspicion, however, that DCM may be a weak genotoxin in mammals.

Data on a possible cytotoxic mechanism are limited and difficult to interpret. Exposure to DCM brought on transient cytotoxic effects in specialized cells of the mouse lung, but the significance of this response to the induction of carcinogenicity is unknown. Further, given that the human lung contains cells biochemically similar to the affected cells in the mouse lung, there is reason to believe that the cytotoxic response in mice could be repeated in humans.

Evidence of an epigenetic mechanism in the liver is still less certain. Data to support the hypothesis that the carcinogenic response in the mouse liver results from increased cell turnover are scant. An S-phase hepatocyte study registered small, variable increases in the incidence of S-phase cells after DCM exposure, but the authors of the study concluded that the biological significance of these changes is unclear.

Current evidence is simply not sufficient either to identify with reasonable certainty the mechanism of action of DCM or to indicate that said mechanism would not be expected in humans. Questions concerning the mechanism may be answered by additional research. The European Council of Chemical Manufacturers Federation (CEFIC) has studies under way to better define the role of the effects of DCM on the Clara cell in the development of mouse tumors (CEFIC, 1987). The results of this research are expected in the near future. The National Institute of Environmental Health Sciences (NIEHS) has planned investigations of the role of cell replication and the pattern of oncogene activation in DCM tumorigenesis. Preliminary results of NIEHS's experiments may be available in 1988.

It does seem likely from the available data that DCM produces a carcinogenic response via glutathione-S-transferase (GST)-mediated metabolism. The correlation between GST activity and the presence of tumors is strong, and there is little evidence to implicate either parent DCM or the alternative mixed function oxidase (MFO) pathway. The level of GST metabolism in the mouse lung and liver has been shown to be high, but the degree to which humans metabolize DCM via the GST pathway is uncertain, at present.

The pharmacokinetic model used by Andersen and Reitz (Andersen et al., 1986, 1987) estimates human GST activity toward DCM to be not too far below the level in mice. The values of Andersen and Reitz may be in conflict with CEFIC's estimates from _in vitro_ measurements of GST activity in human liver samples, however. The CEFIC data indicate an absence of GST activity above the limit-of-detection of the test system. At present, it seems reasonable to assume that humans metabolize DCM via the GST pathway at a rate below that of the mouse, possibly far below.

Neither do the epidemiologic data rule out a risk to humans, although the two studies of humans exposed to DCM in the workplace show no statistically significant increase in deaths from either liver or lung cancer (Ott et al., 1983; Friedlander et al., 1978; Hearne and Friedlander, 1981; Hearne et al., 1986). The study of Kodak filmcasting workers, a particularly well-documented analysis, recorded no statistically significant excess of deaths from any type of cancer, with the possible exception of pancreatic cancer (Hearne et al., 1986). The Kodak cohort had an elevation in pancreatic cancer deaths that is marginally significant if a 5% level of significance is used. The increase in pancreatic cancers is not considered an unequivocal positive response, but it gives some weight to the possibility that a carcinogenic response in humans exposed to DCM could occur at sites others than those found to respond in animals. Further, a quantitative analysis of cancer risks estimated directly from the human data indicates that the results of the Kodak study do not refute the magnitude of risk estimated from the animal data, when animal-based risks are estimated on the basis of internal dose. The epidemiologic data are thus judged to be insufficient to indicate clearly that DCM does not cause cancer in humans.

The EPA believes that the currently available data on mechanism of action, carcinogenic metabolic pathway, and epidemiology do not support a finding of zero cancer risk to humans from DCM exposure. If the risk to humans is not zero, then what is the best estimate of risk?

Principle 26 of the OSTP (1985) states:

> No single mathematical procedure is recognized as the most appropriate for low-dose extrapolation in carcinogenesis. When relevant biological evidence on mechanism of action exists (e.g., pharmacokinetics, target organ dose), the models or procedures employed should be consistent with the evidence.

The OSTP principle is echoed by the EPA Guidelines, which state:

> When pharmacokinetic or metabolism data are available, or when other substantial evidence on the mechanistic aspects of the carcinogenesis process exists, a low-dose extrapolation model other than the linearized multistage procedure might be considered more appropriate on biological grounds.

DCM has been shown to be metabolized in mice by two pathways--one of which, the MFO pathway, is saturated at high doses. Further, the second pathway, which is mediated by GST, appears to be less active in humans than in mice. Andersen et al. have developed a physiologically based pharmacokinetic model which provides a framework for estimating the non-linearity inherent in metabolism by two pathways used to different extents at high and low doses, and, possibly, for incorporating species differences into the risk estimation procedure.

The pharmacokinetic model used by Andersen and Reitz is yet to be fully validated, and some of its parameters, in particular the partition coefficients and the kinetic parameters, are considered to be uncertain. Further, there is debate within the scientific community over the best way to use the model's results in developing estimates of risk. Nonetheless, the pharmacokinetic model provides a means for taking into account metabolic data which would otherwise be ignored in the applied-dose procedure. Despite its uncertainties, the model allows the development of preliminary estimates of risk based on metabolized dose.

A comparison of risk estimates made directly from the human data provided by the Kodak epidemiologic study with risk estimates derived from the results of the pharmacokinetic model used by Andersen and Reitz does not show the animal-based risk estimates to be overestimates, using upper-bound risk estimates from respiratory cancer deaths or using either maximum likelihood estimates or upper-bound estimates from the pancreatic cancer deaths in the Kodak study.

The EPA concludes that the animal evidence of carcinogenicity conforms to the definition for "sufficient" in the EPA Guidelines for Carcinogen Risk Assessment. The epidemiologic studies, while showing no evidence of either liver or lung cancer attributable to DCM, are not sufficient to rule out a risk to humans; the data on deaths from pancreatic cancer give some weight to the possibility that DCM may cause cancer in humans at sites other than those found in animal species. Overall, the epidemiologic data conform to the definition in the Guidelines for "inadequate" insofar as the pancreatic cancer deaths cannot be used to establish a connection between exposure to DCM and human carcinogenicity, yet the possibility of a such a connection cannot be entirely discounted. Thus, DCM meets the Guidelines' criteria for Group B2, probable human carcinogen.

The available body of evidence on the carcinogenic mechanism of action of DCM and on species differences in utilization of the carcinogenic metabolic pathway are not sufficient to support an estimate of zero cancer risk to humans. An evaluation of the weight of evidence does lead to the conclusion, however, that risks should be estimated on the basis of internal dose of the GST metabolite(s). A comparison of the results of the available studies indicates that the GST pathway is the most likely source of the excess tumorigenesis observed in the NTP mouse bioassay.

Taking into account the newest information on pharmacokinetics, mechanism of action, epidemiology, and methods for incorporating this information into the cancer risk assessment, a cancer risk estimate of 4.7×10^{-7} was calculated for continuous inhalation exposure to DCM at a concentration of 1 $\mu g/m^3$.

8 LIST OF ABBREVIATIONS

ADI	acceptable daily intake
ATSDR	Agency for Toxic Substances and Disease Registry
CAA	Clean Air Act
CPSC	Consumer Product Safety Commission
CRS	Congressional Research Service
DCM	dichloromethane; methylene chloride
DHHS	Department of Health and Human Services
EPA	Environmental Protection Agency
FDA	Food and Drug Administration
GST	glutathione-S-transferase
HAD	health assessment document
HRAC	Health/Risk Assessment Committee
IARC	International Agency for Research on Cancer
IRLG	Interagency Regulatory Liaison Group
MFD	mixed function oxidase
NAS	National Academy of Sciences
NOAEL	no-observed-adverse-effect level
NRC	National Research Council
NTP	National Toxicology Program
OAR	Office of Air and Radiation
OHEA	Office of Health and Environmental Assessment
OSHA	Occupational Safety and Health Administration
OSTP	Office of Science and Technology Policy
OTS	Office of Toxic Substances
RCRA	Resource Conservation and Recovery Act
RfD	reference dose
SARA	Superfund Amendments and Reauthorization Act
TSCA	Toxic Substances Control Act
U.S.	United States

9   REFERENCES

Andersen, M. E., Clewell, H. J., III, Gargas, M. L., Smith, F. A., and Reitz, R. H., 1986, Physiologically based pharmacokinetics and the risk assessment process for methylene chloride. Submitted for publication.

Andersen, M. E., Clewell, H. J., III, Gargas, M. L., Smith, F. A., and Reitz, R. H., 1987, Physiologically based pharmacokinetics and the risk assessment process for methylene chloride, _Toxicol. Appl. Pharmacol._, 87:185-205.

CEFIC, 1987 (Mar.), Methylene chloride (dichloromethane): further experimental data, telex from M. Harris, ICI Americas, to J. Hopkins, Office of Toxic Substances, U.S. Environmental Protection Agency, Washington, D.C.

Congressional Research Service, 1983 (Mar.), "A Review of Risk Assessment Methodologies," for the Committee on Science and Technology, U.S. House of Representatives, Washington, D.C., U.S. Government Printing Office.

FDA, 1985, Cosmetics; proposed ban on the use of methylene chloride as an ingredient of aerosol cosmetic products, _Federal Register_, 50(243):51551-51559.

Friedlander, B. R., Hearne, T., and Hall, S., 1978, Epidemiologic investigation of employees chronically exposed to methylene chloride: mortality analysis, _J. Occup. Med._, 20:657-666.

Hearne, F. T., and Friedlander, B. R., 1981, Follow-up of methylene chloride study, _J. Occup. Med._, 23:660.

Hearne, F. T., Grose, F., Pifer, J. W., and Friedlander, B. R., 1986 (June 16), "Methylene Chloride Mortality Study Update," Eastman Kodak Company.

HRAC, 1987, "Technical Analysis of New Methods and Data Regarding Dichloromethane Hazard Assessments," EPA-600/8-87-029, prepared by an interagency work group of the Integrated Chlorinated Solvents Project.

International Agency for Research on Cancer, 1982, "IARC Monographs on the Evaluation of the Carcinogenic Risk of Chemicals to Humans," Supplement 4, Lyon, France.

National Research Council, 1983, "Risk Assessment in the Federal Government: Managing the Process," National Academy Press, Washington, D.C., for the National Academy of Sciences.

NTP, 1985, "NTP Technical Report on the Toxicology and Carcinogenesis Studies of Dichloromethane in F344/N rats and B6C3F$_1$ Mice (Inhalation Studies)," NTP-TR-306.

OSHA, 1986, Occupational safety of methylene chloride, _Federal Register_, 51(226):42257.

OSTP, 1985 (Mar. 14), Chemical carcinogens: a review of the science and its associated principles, February 1985, _Federal Register_, 50(50):10372-10442.

Ott, M. G., Skory, L. K., Holder, B. B., Bronson, J. M., and Williams, P. R., 1983, Health evaluation of employees occupationally exposed to methylene chloride, _Scand. J. Work Environ. Health_, 9(Suppl. 1):8-16.

Preuss, P. W., and Ehrlich, A. M., 1987, The Environmental Protection Agency's risk assessment guidelines, _JAPCA_, 37(7):784-791.

Russo, J., Tay, L. K., and Russo, I. H., 1982, Differentiation of the mammary gland and susceptibility to carcinogenesis, _Breast Cancer Res. Treat._, 2(1):5-73.

U.S. CPSC, 1986, Household products containing methylene chloride: status as hazardous substances, _Federal Register_, 51(161):29778-29809.

U.S. Department of Health and Human Services, 1985 (Apr.), "Risk Assessment and Risk Management of Toxic Substances," prepared by the Committee to Coordinate Environmental and Related Programs.

U.S. EPA, 1985a (Feb.), "Health Assessment Document for Dichloromethane (Methylene Chloride): Final Report," EPA/600/8-82/004F. NTIS PB85-191559.

U.S. EPA, 1985b, Analysis of the applicability of TSCA section 4(f) to methylene chloride, EPA Docket No. OPTS-62045, Federal Register, 50:202126.

U.S. EPA, 1985c (Sept.), "Addendum to the Health Assessment Document for Dichloromethane (Methylene Chloride): Updated Carcinogenicity Assessment of Dichloromethane (Methylene Chloride)," EPA/600/8-82/004FF. NTIS PB86-123742/AS.

U.S. EPA, 1985d, Initiation of regulatory investigation for methylene chloride, Federal Register, 201:42037-42047.

U.S. EPA, 1986a, Guidelines for carcinogen risk assessment, Federal Register, 51:33992-34003.

U.S. EPA, 1986b, Guidelines for estimating exposures, Federal Register, 51:34042-34054.

U.S. EPA, 1986c, Guidelines for mutagenicity risk assessment, Federal Register, 51:34006-34012.

U.S. EPA, 1986d, Guidelines for the health assessment of suspect developmental toxicants, Federal Register, 51:34028-34040.

U.S. EPA, 1986e, Guidelines for the health risk assessment of chemical mixtures, Federal Register, 51:34014-34025.

U.S. EPA, 1987 (July), "Update to the Health Assessment Document and Addendum for Dichloromethane (Methylene Chloride): Pharmacokinetics, Mechanism of Action, and Epidemiology: Review Draft," EPA/600/8-87/030A.

ASSESSMENT CF ECOLOGIC RISKS RELATED TO CHEMICAL EXPOSURE:

METHODS AND STRATEGIES USED IN THE UNITED STATES

J. W. Falco

Office of Environmental Processes and Effects Research
United States Environmental Protection Agency
Washington, D.C.

R. V. Moraski

Office of Health and Environmental Assessment
United States Environmental Protection Agency
Washington, D.C.

1   CURRENT STATUS

At present, the United States has yet to develop government- or agency-wide guidelines for conducting ecologic risk assessments; however, various standard test methods have been developed to provide toxicologic benchmarks. The earliest of these methods measured acute toxicologic effects, but as this field of science progressed, methods to measure chronic effects were also developed. Most recently, research efforts have been directed toward developing test methods that predict chronic and acute toxicologic effects based on results of short-term exposure of organisms during sensitive life stages.

The American Society for Testing and Materials (ASTM) has published many of the earlier methods used in the United States for testing acute and chronic effects. Depending on their scope and level of detail, test procedures are published as ASTM guides, practices, or test methods. A partial compilation of methods developed by the United States Environmental Protection Agency (EPA) or published by ASTM is presented in Tables 5.1, 5.2, and 5.3. These toxicologic methods are grouped according to their use for measuring effects on terrestrial, freshwater, or saltwater organisms.

Ecologic risk assessments performed by the EPA are done primarily by the quotient or ratio method; less frequently used methods include ranking techniques and application factors. The ratio method compares a toxicologic benchmark such as an acute LC_{50} value or a chronic no-effects concentration to a given exposure concentration to provide an estimate of risk. As the ratio for a given species approaches a critical value, a high risk is inferred. Exposures of varying intensities and data on ecologic effects are evaluated depending on the purpose of the assessment and the legal requirements that specify the scope of the assessment.

Table 5.1. Methods for Estimating Toxicologic Effects on Terrestrial
Species and Birds of Exposure to Potentially Toxic Chemicals

Method	Reference
Standard Method for Effective Bird Control	ASTM, 1986a
Standard Method for Percutaneous Toxicity	ASTM, 1986b
Standard Method for Subchronic Dermal Toxicity	ASTM, 1986c
Standard Practice for Determining Acute Oral LD_{50} for Testing Vertebrate Control Agents	ASTM, 1986d

Table 5.2. Methods for Estimating Toxicologic Effects on Freshwater
Organisms of Exposure to Potentially Toxic Chemicals

Method	Reference
Methods for Acute Tests with Fish, Macroinvertebrates, and Amphibians	U.S. EPA, 1975 ASTM, 1986e
Method for Aquatic Multiple Species Toxicant Testing	Phipps and Holcombe, 1985
Methods for Conducting Effect Studies on Snail (Aplexa hypnorum) Embryo through Adult Exposures	Holcombe et al., 1984
Standard Practice for Conducting Static Acute Toxicity Tests on Wastewaters with Daphnia	ASTM, 1986f
Standard Guide for Assessing the Hazards of a Material to Aquatic Organisms and their Uses	ASTM, 1986g

2 FUTURE DIRECTIONS

State-of-the-art assessment of risk to the ecosystem is still evolving. Although the single-species tests listed in Tables 5.1, 5.2, and 5.3
have provided valuable information for the assessment of ecologic risk, it
is necessary to focus on ecosystems-level tests and analyses. The increasing availability of predictive models makes assessment of risk to the
environment, rather than simply to a single species, more possible.

Predicting an ecosystem's response to pollutant stress is difficult
because of the large number of dependent and independent variables consti-
tuting and inherent to a natural ecosystem. These include population-
level factors such as density, immigration, growth, and mortality, and
community-level factors such as diversity, relative dominance, trophic
structure, and distribution.

Table 5.3. Methods for Estimating Toxicologic Effects on Saltwater
Organisms of Exposure to Potentially Toxic Chemicals

Method	Reference
Sea Urchin DNA-Based Embryo Growth Toxicity Test	Jackim and Nacci, 1984, 1986
Sea Urchin Sperm Cell Toxicity Test	Dinnel et al., 1983 Beckman, 1982
Bacterial Toxicity Test (Microtox®)	Beckman, 1982 Nacci, 1986
Phorocephalid Amphipod Bioassay	Swartz, 1985
Rhodophyta Life Stages Toxicity Test	Steele and Thursby, 1983
Atherinid Fish Early Life Stage Toxicity Test	Goodman et al., 1985a
Sheepshead Minnows Life Cycle Toxicity Test	Hansen and Parrish, 1977
Cytogenetic Model for Marine Genetic Toxicology	Pesch et al., 1981
Method to Measure Scope for Growth Index for Blue Mussels	Nelson et al., 1985
Method for Measuring AEC as a Test for Stress in Mussels	Zaroogian et al., 1982
Method for Measuring Bioaccumulation of Chemicals in Mussels and Polychaetes	Lake et al., 1985
Method for Measuring Sister Chromatid Exchange in Marine Polychaetes	Pesch et al., 1984
System for Preliminary Evaluation of Infectivity and Pathogenesis of Insect Virus in Shrimp	Couch and Martin, 1984
Tidewater Silversides (Menidia peninsulae) Early Life Stage Toxicity Test	Goodman et al., 1983
Early Life Stage Toxicity Test Using California Grunion	Goodman et al., 1985b

There are ways to simplify the complex structure of an ecosystem. For example, determination and analysis of a key species may facilitate prediction of the effects of pollutant stress on dependent species. In addition, knowledge of physico-chemical parameters of pollutants may make possible the analysis of pollutant fate and transport (see, for example, Chapter 7 of this book). Nevertheless, ecosystem-level analysis is an inherently complex undertaking. Ecosystems may modify the fate and transport of environmental pollutants. In aquatic systems, for example, microbial methylation of mercury may occur with a subsequent accumulation of neurotoxic methylmercury in fish. A number of the factors to be included in any discussion of ecological risk assessment are discussed below.*

2.1 End Points

A variety of ecotoxicologic end points have been proposed to assess the effects of pollutants on ecologic systems. Potential end points occur at the level of the individual organism, the population, and the ecosystem. In general, end points at lower levels of organization (organism or suborganism levels) have been used more widely because they are simpler, are more rapidly and inexpensively assessed, and are most useful in determining the mechanisms of toxicologic effects. End points at the population or ecosystem levels of organization are more complex and difficult to interpret but are probably ecologically more realistic, because they incorporate the complexity of interactions among organisms and between organisms and their abiotic environment. A major, unresolved question is the extent to which end points at lower levels of organization can be used to predict pollutant impacts at higher levels of organization.

2.1.1 Ecosystem Structure End Points. Measures of ecosystem structure can provide important data for ecosystem risk assessments. Structural changes in stressed ecologic communities may be visualized as an information network reflecting environmental conditions but not demonstrating the external mechanisms or internal interactions that brought about a reorganization in species composition or dominance patterns.

Structural end points such as abundance (McNaughton and Wolf, 1973) and biomass (Clapham, 1973) of communities provide relatively simple, gross measurements of ecosystem stress. Species richness has been shown to be sensitive to the level of stress and can provide a partial picture of changes in community composition which accompany stress (McNaughton and Wolf, 1973).

Combined numerical indices such as similarity (Hellawell, 1977) and ordination (Odum, 1971) measures may be used to track changes in community structure which occur as pollutant concentrations change. Although diversity indices (Odum, 1971; Herricks and Cairns, 1982) have been used widely in hazard assessment studies (see, for example, Chapter 10 of this book), these integrated measures are often insensitive to stress and provide data that are difficult to interpret (Hellawell, 1977). The use of numerical indices exclusive of the biologic data from which they are calculated should be discouraged.

2.1.2 Ecosystem Function End Points. The analysis of functional response end points can provide data on energy flow and nutrient cycles. The functional capability of the ecosystem is, in fact, the ultimate criterion of ecologic success. The effective use of end points in describing impacts is dependent on a theoretical and practical knowledge of ecosystems for proper interpretation, and on collection of sufficient baseline data to establish normal process rates. A history of measuring functional response variables will be necessary to establish threshold values for unacceptable reductions in functional capability.

Primary productivity (McNaughton and Wolf, 1973) provides the energy for the base of the food web. This process has been shown to be sensitive to a variety of pollutants and other forms of stress. Reductions in primary productivity, which are of substantial magnitude and long duration,

---

*Some of the discussion that follows is taken from material submitted by Technical Resources, Inc., Rockville, MD, for work performed under EPA contract no. 68-02-4199.

are unquestionably detrimental to energy processing in exposed ecosystems.

Disruptions in material cycles such as the nitrogen cycle (Westman, 1985; Cook, 1984) can be critical if the effects on cycling processes indirectly inhibit ecosystem production. Material cycles can be upset by pollutant inhibition of the decomposition process, interference with the functional links in specific nutrient cycles, or disruption of nutrient conservation mechanisms. Effects on decomposition can be measured in terrestrial and aquatic ecosystems, and changes in decomposition rate and completeness of mineralization can be related to the level of pollution stress. At present, few data are available on the long-term impacts of reduced decomposition on ecosystem production.

Specific nutrient cycling processes are key to the production efficiency of ecosystems. Identification of the critical cycles in specific ecosystems will be necessary for the selection of appropriate monitoring points.

Nutrient conservation is exceedingly important in terrestrial ecosystems. Evidence of excessive leaching of essential nutrients is a sign of stress. Leaching loss of nutrients has been correlated with reduced nutrient availability in the plant-root zone and reduced plant growth in nutrient-deficient soil (Jackson et al., 1979).

A problem in the use of ecosystem function end points is their relative insensitivity to ecosystem structure. Shifts in species composition to more pollutant-resistant species may or may not result in changes in such functional processes as productivity or nutrient cycling. Thus, an assessment of pollutant effects at the ecosystem level should include both structure and function end points.

Because the factors controlling ecosystem structure and function are numerous and poorly understood, it is difficult to distinguish ecosystem-level effects of pollutants from naturally occurring processes. Many of the ecosystem-level end points depend on the questionable assumption that unpolluted ecosystems are at a stable, undisturbed state.

2.1.3 Population-Level End Points. At the population level, stress response may be monitored in terms of changes in the abundance, distribution, age structure, or gene makeup of exposed populations. The first three end points can be related quite clearly to the overall success of the exposed population. Changes in the gene pool may be related to future adaptability of the population to similar types of stress.

Also in question is the selection of an appropriate population or populations to be monitored in an impact assessment. Quite clearly, monitoring effects on commercially or aesthetically valuable species is important for predicting impacts on those species. More valuable for predicting higher level impacts are population response data on representative and ecologically important species within exposed communities. Included within this category are keystone species that strongly influence the structure of the communities or the functioning of the ecosystem. If there is interest in extrapolating population response to predict ecosystem-level impacts, emphasis should be placed on gathering data on populations from major functional groups, including primary producers, primary, secondary, and tertiary consumers, and decomposers.

A problem in using population-level end points as indicators of the effects of pollution is that the numerous factors regulating population structure are, as yet, poorly understood. This makes it difficult to discriminate pollutant effects from naturally occurring processes. As

65

population structure is influenced by interactions among population members, with other populations, and with the abiotic environment, it becomes necessary to examine effects of pollutants at the ecosystem level.

2.1.4 Physiologic End Points. The physiologic end points most closely related to individual fitness are acute mortality, growth and development, and reproductive success. Acute lethality testing such as LD_{50} or LC_{50} determination is widely used to provide minimal estimates of toxicity. However, such testing is not sufficiently sensitive to assess sublethal or chronic effects that occur at lower toxicant concentrations and that may be of considerable ecologic importance.

Biochemical response end points may provide information on mechanisms of toxic action. Since biochemical processes are in general particularly sensitive to pollutants, biochemical response end points may provide early warning of potential impacts on the individual. However, most biochemical processes also respond to conditions other than pollutant stress, and the response of these end points may be adjusted as an individual acclimates to a stress. Correlations between biochemical response end points and individual success need to be established to enhance the value of these sensitive end points as predictors of higher level impacts.

Osmoregulatory activity is an appropriate end point for assessing impacts on certain freshwater and estuarine fish and invertebrates. Again, the ability of individual organisms to acclimate to osmoregulatory stress must be considered in interpreting osmoregulatory response data. Musculoskeletal end points have also been used to monitor stress responses in fish. Correlations need to be established between abnormalities and the ecologic success of deformed fish.

Respiratory activity has been used as a response end point for a number of species. However, it is difficult to generalize about the patterns of respiratory response to stress. Respiration rates may be elevated or inhibited by pollutants, and ventilation rates in exposed individuals may adjust as acclimation occurs.

Behavioral alterations are appropriate end points for impact assessments if the alterations act either to protect the individual from harm, as in avoidance behavior, or to make the individual more vulnerable to the stress, as in the loss of antipredator behavior. Although behavioral responses are not easy to demonstrate in the laboratory or in the field, these end points, if demonstrated, may be easily extrapolated to predict potential population-level effects.

Genotoxicity and carcinogenicity are end points that provide early warning of stress. Data must be gathered on the natural incidence of mutations and tumors to aid in interpreting the importance of chemically induced mutation and tumor incidence rates.

End points measuring growth, development, and reproductive success of individuals are of most obvious utility in predicting population-level impacts. Because these end points are directly related to population success, their use is recommended in impact studies where single-species test data are extrapolated to predict population-level impacts. These end points have been used less frequently because of the time and expense required to conduct full-life-cycle chronic toxicity tests. However, the more frequently used short-term physiologic and biochemical end points cannot be recommended until their relationships to organismal growth and reproductive success are determined.

A number of studies (Babich and Stotzky, 1980; Lighthart, 1980;

66

Reinert and Spurr, 1972; Miles and Parker, 1980) have documented interactions between effects of pollutants and abiotic and biotic factors in the environment. These studies illustrate the inadequacy of using laboratory single-species, single-factor testing to estimate all ecologic effects of contaminants, and they point to the necessity of relating ecotoxicologic effects on individual organisms to population- and ecosystem-level effects of pollutants.

2.1.5 End Points and Ecological Risk Assessment. A multilevel ecologic risk assessment, which makes use of a combination of organism, population, and ecosystem-level end points, provides the most effective approach to examining ecosystem stress. Multilevel testing would both enhance the sensitivity of a risk assessment and broaden its scope to include more complex levels of ecologic organization. In contrast, the traditional approach of using only single-species testing is generally inadequate to account for pollutant-induced effects on the complex organization of an ecosystem. Single-species measures can be greatly enhanced by the use of population and ecosystem-level end points.

The precise choice of end points for use in an ecologic risk assessment should be made on a case-by-case basis, depending on both the ecosystem being tested and the nature of the pollutant stressor. Various population- and ecosystem-level end points are potential choices. Many of these end points are readily measurable and are highly sensitive to low levels of pollutant stress. Still inadequate, however, are field data documenting the usefulness of population- and ecosystem-level end points in ecosystem toxicity studies. Future research in this area would facilitate the development of the multilevel risk assessment approach.

2.2 Choice of Species

The choice of species to study in an ecosystem is also important; typically, the focus is endangered or sensitive species. The selection of ecosystem media and interaction of pollutants within these media further complicate ecosystem assessment. Ecosystems incorporate processes that operate on diverse spatial, structural, or temporal scales. The enmeshing of these variables presents difficulties in calculating the effects of localized versus general processes and in integrating key factors such as primary production with seasonal climatic changes and geochemical cycling.

Intermediate between full field tests and single-species laboratory bioassays are microcosm and mesocosm studies. Microcosms and intermediate mesocosms are isolated parts of a naturally occurring system that can be modified to duplicate many features of an intact ecosystem. Microcosms have many limitations including those of spatial scale, number of organisms, species diversity, and physical controlling variables.

2.3 Use of Models

Various models can be used to evaluate ecosystem risk. These include models of fate, transport, exposure, and effects as well as integrative models (see, for example, Chapter 7 of this book). Other ecosystem models focus on population density, food chains, bioenergetics, and toxicokinetics. The diverse models for both individual species and population groups have advantages and disadvantages that must be defined and tailored to specific circumstances. This diversity provides for a wide variety of approaches that can be applied to the problems encountered in ecologic risk assessment.

Succession models take on a wide range of mathematical forms. Their richness in formulation originates to an extent in the specific objectives and training of the model designer. These differences frequently imply different theoretical constructs as to what is important in the functioning of a given ecosystem. In this sense, the models represent a complex set of a priori hypotheses about the function and behavior of ecosystems.

Thus it is essential to use an orderly, justifiable process in developing and selecting an appropriate ecosystem model. Refining and improving available models are critical aspects of developing precise models for each particular situation in nature.

One of the most pressing needs in environmental management is for evaluation of models used routinely in assessment of environmental exposure and impact. Many models are being used for situations where they are of questionable validity. In particular, models should not be applied to environments outside the range for which they have been calibrated and tested. Although this may seem obvious, oftentimes models are used to predict impacts for changing conditions that are appreciably different from those for which the models were originally developed and calibrated.

With regard to population modeling, although current work on mathematical models of individuals and populations shows great promise, insufficient scrutiny to date has precluded a general consensus on approach. Most rudimentary population models have been developed from a retrospective viewpoint with biologic data, but few biologic principles, as a focus. Such models are useful for risk assessment only from a qualitative perspective. However, discrete age- or stage-structured population models offer a well-developed theory and a reasonable computational scheme.

Another significant problem in current ecosystem risk assessment is the paucity of toxic effects models and of predictive methods. Validation of methods to predict ecosystem response is difficult because of the absence of empirical data in many areas. Generation of data, development of models, and validation of methods are current projects of the EPA's Office of Environmental Processes and Effects Research.

2.4 <u>Resilience and Recovery</u>

Factors that influence recovery of an ecosystem from environmental stress include severity of the stress, reversibility of effects, rate and effectiveness of stress removal, frequency and duration of ecosystem disturbance, resilience of ecosystem structure and function, extent of alteration, compensatory interaction of multiple species, kinetic balance of the system, complexity of the system, temporal and spatial variability, availability of regenerating units, and rate of reestablishment of the biologic and physical habitat.

The resilience of ecologic systems and their resistance to natural and anthropogenic forms of disturbance have been measured in field and laboratory studies. Of necessity, these studies have been of long duration. In most cases, natural ecosystems have not been shown to be displaced to the extent that recovery is not possible when the disturbance abates (Sheehan, 1984). The availability of colonizers to the disturbed ecosystem and the existence of biogeochemical feedback loops are cited as factors important to the rapid recovery of disturbed ecosystems. Functional redundancy of species is cited as important to the resistance of ecosystems to disturbance. Loss of individual populations may not in itself be an adequate measure of the stability of the structure and function of a disturbed ecosystem.

Resistance and resilience are appropriate response variables for impact assessment. These stability criteria represent integrated measures of system breakdown and recovery. In future studies, measurement of amplitude will be essential to establish measurable threshold levels of disturbance which may be indicative of permanent changes in stressed ecosystems.

2.5 Development of Surrogate Systems

One potential tool for quantification of health and environmental effects is the classification of well-described ecosystem types for use as surrogates for candidate ecosystems. Establishment of such surrogate systems would simplify evaluation without excessive loss of accuracy.

Screening methodologies can be used for certain classes of chemicals to predict chemical persistence and the potential for bioaccumulation based on physico-chemical properties and quantitative structure-activity relationships. On the basis of screening results, biologic testing may be recommended. Further assessment with integrated exposure/effects models, microcosm/mesocosm experiments, or field studies are logical extensions of such screening efforts.

2.6 Uncertainty

The uncertainty associated with ecosystem risk analysis can arise from a variety of different sources and in a number of different ways that affect the calculation of risk. Perhaps foremost among these is that the response of ecosystems, or their components, to anthropogenic stress involves numerous factors. Each of these factors incorporates physical or biologic mechanisms that in turn vary in degree of scientific characterization, availability of data sets, and sources and levels of uncertainty. Thus, natural complexity and stochasticity contribute to the uncertainty associated with models. Because ecosystem risk analysis typically includes a mathematical or statistical model, lack of correspondence between the model and the modeled ecosystem leads to model error. Errors in parameter estimates resulting from experimental measurement error, approximation and extrapolation of experimental results, and solution techniques also contribute uncertainties to ecologic risk assessment (see also Chapter 2 of this book).

2.7 Integrated Strategy

An integrated strategy including single-species bioassays, microcosm and mesocosm experiments, and models for exposure and toxic effects allows an estimate of the biologic effects of a physical or chemical stress. If model parameters can be obtained from actual test data, then model accuracy can be improved by stepwise calibration of models to microcosms and mesocosms. Thus, variable natural conditions can be represented more realistically and ecologic risk estimated with fewer uncertainties.

3. ECOLOGICAL RISK ASSESSMENT GUIDELINES

Ecosystem risk assessment appears to be a feasible undertaking when its limitations are clearly delineated. The EPA's Office of Health and Environmental Assessment is currently developing guidelines that will provide a general approach for conducting ecologic risk assessments. The guidelines will help the assessor identify the pathways and mechanisms by which chemicals reach nonhuman populations; from an understanding of the chemical effects, the assessor will then develop an assessment of risk. The guidelines will help the assessor to determine which aspects of the ecosystem to emphasize and whether available data are adequate to estimate exposure and effects of concern. The risk manager will then have a basis for deciding what constitutes an unreasonable ecologic risk.

The guidelines will discuss the fundamental principles governing the response of the environment to stress not only for individual organisms but also for populations of organisms. Discussion will include current, acceptable methods for testing effects of pollutants on single and multiple species; biologic, molecular, and physiologic indicators of toxicity; pharmacodynamic and environmental mechanisms of toxic effects; and ecosystem-level functions such as nutrient processing, productivity, and diversity as indicators of toxic effects.

The guidelines will allow the risk assessor to consider the following questions when developing an ecologic exposure assessment: how does the ecosystem modify the fate and transport of the toxicant; how is the contaminant distributed within the ecosystem; what are the residence times; what are the sites of retention or deposition; what fate and transport models would be helpful in determining environmental concentrations; is it possible to combine bioassay data with models, microcosm studies, and field-study methods to determine transport, fate, and potential exposure; what is known about the natural dynamics of the ecosystem; what is the extent and duration of exposure for the biota; does the ecosystem recover from the stress, and how is recovery measured; and are any sensitive or endangered species, or species at vulnerable life stages, present in the ecosystem being studied?

The guidelines will help the assessor develop an ecologic hazard assessment. Elements that may contribute to a hazard assessment include factors that affect the toxicity of the chemical; parent, metabolite, or degradation products responsible for the toxic effects; selection of models based on an intent to study effects on individual species, certain population groups, or the ecosystem as a whole; comparison of changes that occur in the environment in the absence of stress with changes that occur in the presence of stress; identification of ecologically important species; possible synergistic mechanisms; intent to study acute or chronic effects or both; appropriateness of laboratory-to-field extrapolations; availability of appropriate benchmark compounds; availability of applicable retrospective cases; appropriateness of an approach involving a surrogate species or ecosystem; and presence of ecologic indicators in the air, water, soil, and ecosystems.

The guidelines will help the assessor choose the best monitoring system for the assessment, design the sampling plan, determine the role of models, and decide whether a tiered testing approach is necessary to predict higher-order effects.

Finally, the guidelines will help the assessor to develop a risk assessment that integrates the exposure and hazard assessments. Key elements that should be considered for inclusion in the risk assessment include selection of end points, description of the reference environment, identification of sources, assessment of exposure and effects, integrated risk assessment, and evaluation of uncertainty.

4   REFERENCES

ASTM, 1986a, Standard methods for developing effective bird control chemicals, in: "Annual Book of ASTM Standards," Vol. 11.04, Philadelphia.

ASTM, 1986b, Standard test method for mammalian acute percutaneous toxicity, in: "Annual Book of ASTM Standards," Vol. 11.04, Philadelphia.

ASTM, 1986c, Standard method for determining subchronic dermal toxicity, in: "Annual Book of ASTM Standards," Vol. 11.04, Philadelphia.

ASTM, 1986d, Standard practice for determining acute oral LD$_{50}$ for testing vertebrate control agents, in: "Annual Book of ASTM Standards," Vol. 11.04, Philadelphia.

ASTM, 1986e, Standard practice for conducting acute toxicity tests with fishes, macroinvertebrates and amphibians, in: "Annual Book of ASTM Standards," Vol. 11.04, Philadelphia.

ASTM, 1986f, Standard practice for conducting static acute toxicity tests on wastewaters with Daphnia, in: "Annual Book of ASTM Standards," Vol. 11.04, Philadelphia.

ASTM, 1986g, Standard guide for assessing the hazard of a material to aquatic organisms and their uses, in: "Annual Book of ASTM Standards," Vol. 10.04, Philadelphia.

Babich, H., and Stotzky, G., 1980, Physiochemical factors that affect the toxicity of heavy metals in microbes in aquatic habitats, in: "Aquatic Microbial Ecology," Proc. Conf. Amer. Soc. Microbiology, R. R. Colwell and J. Foster, eds., Maryland Sea Grant Publ. University of Maryland, College Park, MD.

Beckman Instruments Inc., 1982, "Microtox® System Operating Manual," Fullerton, CA.

Clapham, W. B., 1973, "Natural Ecosystems," Case Western Reserve University, Cleveland, OH.

Cook, R. B., 1984, Man and the biogeochemical cycles: interacting with the elements, Environment, 26(7):11-40.

Couch, J. A., and Martin, S. M., 1984, A simple system for the preliminary evaluation of infectivity and pathogenesis of insect virus in a nontarget estuarine shrimp, J. Invert. Pathol., 43:351-357.

Dinnel, P., Stoher, Q., Letourneau, M., Roberts, W., Felton, S., and Nakatani, R., 1983, "Methodology and Validation of a Sperm Cell Toxicity Test for Testing Topic Substances in Marine Waters," Final Report, Grant R/Tox. FRI-UW-83, University of Washington Sea Grant Program in cooperation with U.S. Environmental Protection Agency.

Goodman, L. R., Hansen, D. J., Middaugh, D. P., Cripe, G. M., and Moore, J. C., 1985a, "Method for Early Life-Stage Toxicity Tests Using Three Atherinid Fishes and Results with Chlorpyrifos, Aquatic Toxicology and Hazard Assessment: Seventh Symposium," American Society for Testing and Materials Special Technical Publication 854, R. D. Cardwell, R. Purdy, and R. C. Bahner, eds., American Society for Testing and Materials, Philadelphia.

Goodman, L. R., Hansen, D. J., Cripe, B. M., Middaugh, D. P., and Moore, J.C., 1985b, A new early life-stage toxicity test using the California grunion (Leuresthes tenuis) and results with chlorpyrifos, Ecotoxicol. Environ. Safety, 10:12-21.

Goodman, L. R., Middaugh, D. P., Hansen, D. J., Higdon, P. K., and Cripe, G. M., 1983, Early life-stage toxicity test with tidewater silversides (Menidia peninsulae) and chlorine-produced oxidants, Environ. Toxicol. Chem., 2:337-342.

Hansen, D. J., and Parrish, P. R., 1977, "Suitability of Sheepshead Minnows (Cyprinodon variegatus) for Life-Cycle Toxicity Tests, Aquatic Toxicology and Hazard Evaluation," American Society for Testing and Materials Special Technical Publication 634, F. L. Mayer and J. L. Hamelink, eds., American Society for Testing and Materials, Philadelphia.

Hellawell, J. M., 1977, Change in natural and managed ecosystems: detection, measurement and assessment, Proc. R. Soc. Lond. B. Biol. Sci., 197:31-56.

Herricks, E. E., and Cairns, Jr., J., 1982, Biological monitoring. Part III--Receiving methodology based on community structure, Water Res., 16:141-153.

Holcombe, G. W., Phipps, G. L., and Marier, J. W., 1984, Methods for conducting snail (Aplexa hypnorum) embryo through adult exposures: effects of cadmium and reduced pH levels, Arch. Environ. Contam. Toxicol., 13:627-634.

Jackim, E., and Nacci, D., 1984, A rapid aquatic toxicity assay utilizing labeled thymidine incorporation in sea urchin embryos, _Environ. Toxicol. Chem._, 3:631-636.

Jackim, E., and Nacci, D., 1986, Improved sea urchin DNA-based embryo growth toxicity test, _Environ. Toxicol. Chem._, 5:561-565.

Jackson, D. R., Ausmus, D. S., and Levin, M., 1979, Effects of arsenic on the nutrient dynamics of grassland microcosms and field plots, _Water Air Soil Pollut._, 10:13-18.

Lake, J., Hoffman, G. L., and Schimmel, S. C., 1985, "Bioaccumulation of Contaminants from Black Rock Harbor Dredged Material by Mussels and Polychaetes," Technical Report D-85-2, Department of the Army, U.S. Army Corps of Engineers, Washington, D.C.

Lighthart, B., 1980, Effects of certain cadmium species on pure and litter populations of microorganisms, _Antonie Van Leeuwenhoek_, 46:161-167.

McNaughton, S. J., and Wolf, L. L., 1973, "General Ecology," Holt, Rinehart and Winston, New York.

Miles, L. J., and Parker, G. R., 1980, Effects of cadmium and a one-time drought stress on survival, growth, and yield of native plant species, _J. Environ. Qual._, 9:278-282.

Nacci, D., 1986, Comparative evaluation of three rapid marine toxicity tests: sea urchin early embryo growth test, sea urchin sperm cell toxicity test and microtox, _Environ. Toxicol. Chem._, 5:523.

Nelson, W. G., Black, D., and Phelps, D., 1985 (September), "Utility of the Scope for Growth Index to Assess the Physiological Impact of Black Rock Harbor Suspended Sediment on the Blue Mussel, _Mytilus edulis_: A Laboratory Evaluation," Technical Report D-85-6, Department of the Army, U.S. Army Corps of Engineers, Washington, D.C.

Odum, E. P., 1971, "Fundamentals of Ecology," W. B. Saunders Co., Philadelphia.

Pesch, G., Heltske, J., and Mueller, C., 1984, A statistical analysis of _Neanthes arenaceodentata_ sister chromatid exchange data, _in_: "Sister Chromatid Exchanges," R. R. Tice and A. Hollaender, eds., Plenum Publishing Corp., New York.

Pesch, G. G., Pesch, C. E., and Malcolm, A. R., 1981, _Neanthes arenaceodentata_, a cytogenetic model for marine genetic toxicology, _Aquatic Toxicol._, 1:301-311.

Phipps, G. L., and Holcombe, G. W., 1985, A method for aquatic multiple species toxicant testing: acute toxicity of 10 chemicals to 5 vertebrates and 2 invertebrates, _Environ. Pollu._, 38:141-157.

Reinert, R. A., and Spurr, H. W., 1972, Differential effects of fungicides on ozone injury and brown spot disease of tobacco, _J. Environ. Qual._, 1:450-452.

Sheehan, P. J., 1984, Functional changes in the ecosystem, _in_: "Effects of Pollutants at the Ecosystem Level," P. J. Sheehan, D. R. Miller, G. C. Butler, and P. Bordeau, eds., John Wiley and Sons, New York.

Steele, R. J., and Thursby, G. B., 1983, "A Toxicity Test Using Life Stages of _Champia parveila_ (_Rhodophyta_), Aquatic Toxicology and Hazard Assessment: Sixth Symposium," American Society for Testing and Materials Special Technical Publication 802, W. E. Bishop, R. D. Cardwell, and B. B. Heidolph, eds., American Society for Testing and Materials, Philadelphia.

Swartz, R. C., 1985, "Phorocephalid Amphipod Bioassay for Marine Sediment Toxicity, Aquatic Toxicology and Hazard Assessment: Seventh Symposium," American Society for Testing and Materials Special Technical Publication 854, R. D. Cardwell, R. Purdy, and R. C. Bahner, eds., American Society for Testing and Materials, Philadelphia.

U.S. Environmental Protection Agency: Committee on Methods for Toxicity Tests with Aquatic Organisms, 1975, "Methods for Acute Tests with Fish, Macroinvertebrates and Amphibians," Ecological Research Series, EPA-600/375-009, NTIS No. PB 242105, National Technical Information Service, Springfield, VA.

Westman, W. E., 1985, "Ecology, Impact Assessment, and Environmental Planning," John Wiley and Sons, New York.

Zaroogian, G. E., Gentile, J. H., Heltske, J. F., Johnson, M., and Ivanovici, A. M., 1982, Application of adenine nucleotide measurements for the evaluation of stress in _Mytilus edulis_ and _Crossotrea virginia_, _Comp. Biochem. Physiol._, 71B(4):643-649.

ATTITUDES TOWARD RISK-BENEFIT ANALYSIS FOR MANAGING EFFECTS OF CHEMICAL
EXPOSURES

J. L. Regens

Institute of Natural Resources
University of Georgia
Athens, GA

1   INTRODUCTION

Environmental policymaking has become more dependent on the use of
formal, quantitative risk-benefit analysis as a basis for establishing
regulations.   This is largely because of increasing emphasis on the
prevention of adverse human health effects attributable to exposures to
toxic chemicals.   For example, risk assessment helps to set priorities
for the regulation of the very large number of chemicals which are of
potential concern by ranking those chemicals in terms of their possible
carcinogenic or other health hazards (see Ames et al., 1987).   Moreover,
when linked to benefit-cost studies, risk assessment can help direct
limited societal resources to prevent or mitigate the most significant
risks (see also Chapters 4 and 5 of this volume; Rycroft et al., forthcoming; Russell and Gruber, 1987; Lave, 1987).

Risk-benefit analysis, therefore, aids regulatory decisionmakers in
setting priorities and adjusting regulations to the degree and distribution of risk being controlled.   Risk-benefit analysis also can be helpful
as a basis for making site-specific judgments.   For example, risk-benefit
analysis makes it possible to consider systematically information about
the nature of the pollutant, the sensitivity of the environmental
setting, the availability of control technologies, and the net benefits
of control options.   As a result, the application of formal analytical
tools frequently is advocated as a means for rationalizing environmental
policy formulation and implementation (Douglas, 1985; Barker, 1984;
Ruckelshaus, 1983).

Although risk-benefit analysis can be a valuable tool for incorporating scientific evidence and economic information into the legal and adminstrative process, its use as a basis for formulating and implementing environmental policy to manage the adverse effects of chemical exposures remains controversial (Rycroft et al., 1987; Hoos, 1983; Swartzman et al., 1982). Because of this, it is important to delineate empirically the factors influencing the use of risk assessment and economic analysis by regulatory agencies. This research, by focusing on attitudes in the U.S. Environmental Protection Agency (EPA) toward using risk assessment and economic analysis, provides such an appraisal of regulatory agency receptivity to the use of risk-benefit analysis for managing human health and ecological effects of chemical risks.

2   DATA AND METHODS

Data for this study were obtained through a series of interviews conducted with 34 senior officials of the U.S. Environmental Protection Agency whose responsibilities involved assessing risks or developing policies to prevent or mitigate environmental damages. Since the EPA has some internal discretion in determining the role of technical analysis in environmental regulation, the interviews focused on senior staff and managerial personnel at its Washington, DC headquarters. The interviews were conducted using a face-to-face format. Each of the EPA officials was asked a series of closed-ended questions in order to collect information about their personal backgrounds as well as their attitudes towards key aspects of risk-benefit analysis (see Appendix). In all cases, this group of EPA officials were assured that their responses would only be presented in summary statistics, all individuals were guaranteed confidentiality, and their participation in the study was completely voluntary.

By interviewing employees who were either members of the Senior Executive Service (SES) or at the GM-15 level which is immediately below the SES, it was possible to gain information about the attitudes of individuals within the EPA who are responsible on a day-to-day basis for regulatory decisionmaking. The overwhelming majority of interviewees were males (73.5%), and the median age for the individuals who participated in this study was 42.8 years (s = 6.7 years). Interviewees tended to have earned their highest degree in the physical sciences or engineering (47.1%). This is not surprising because of the technical complexity of the problems with which the EPA has to deal. The strong emphasis on scientific and technical backgrounds is further underscored by the fact that almost 65 percent of the interviewees had obtained their highest degrees in the physical sciences, engineering or environmental sciences. Educational background was indicated by the field in which an individual received his/her highest degree. Degrees were coded into the following categories: (1) social sciences/liberal arts (26.5%) which include social sciences, economics, business, management, liberal arts and humanities; (2) environmental sciences (17.6%) which include biology, ecology, environmental science; and one individual with a biomedical background; (3) physical sciences/engineering (47.1%) which include physical sciences, mathematics, statistics, and engineering; and (4) law (8.8%).

Despite the fact that a majority of these senior managers have a background in the sciences, because research and development is not EPA's primary mission, most of them worked in policy support or program offices. The interviewees indicated that their job responsibilities primarily involved the translation of scientific information or the formulation and implementation of policy options to address environmental issues such as managing the effects of chemical exposures. Job responsibilities were coded into three categories: (1) scientific research (17.6%); involving the conduct and/or supervision of research (2) information translation (52.9%); which involves the development of policy from research or the translation and interpretation of scientific information and (3) policy formulation/implementation (29.4%) which involves working directly on policy issues, lobbying, litigation, and legal research.

In order to assess senior EPA officials' potential receptivity to the use of risk-benefit analysis for managing adverse health effects due to exposure to toxic chemicals in the environment, this study focuses on four key aspects of their attitudes towards such formal problem-solving techniques. The first consists of their willingness to support the use of formal risk analysis in environmental policymaking. The second is their support for using benefit-cost analysis in environmental policy-making. Third, whether they think that society must attempt to place an economic value on human life in order to allocate scarce resources? And, finally, each individual's preferred method for the distribution of risks and benefits in developing environmental regulations.

Examining the perspectives of this particular set of individuals towards those questions is important. Such information is useful to have because the EPA is a pioneer in the application of risk assessment and benefit-cost analysis to environmental management. It can help to clarify substantially attitudes among key officials at the EPA towards the use of risk-benefit analysis for managing the adverse human health and ecological effects of chemicals exposures. As a result, the interview data offers insights into the perceptions of individuals who have substantial experience using risk-benefit analysis for managing effects of chemicals exposures, especially because some disagreement surrounds the use of such tools as a basis for environmental regulation. Judgments about whether a hazard is voluntary or involuntary as well as due to environmental or "lifestyle" factors may influence the choice of regulatory options adopted to prevent or mitigate risks. For example, debate exists within the scientific community over the proportion of cancer deaths due to particular causes. Environmental scientists typically claim that 40 to 80 percent are attributable to chemical hazards while biomedical researchers often view lifestyle factors such as smoking or intake of fat as more being significant causally. Perhaps the seeming disagreement within the scientific community is mainly a matter of definition. That is, smoking is a lifestyle factor but it is also a chemical factor. Technical subtleties aside, small variations in definition and assumptions can produce large differences in risk estimates and resultant benefits valuation (see Epstein and Swartz, 1981; Peto, 1980; Wynder and Gori, 1977).

The present appraisal permits an empirical evaluation of the relationship between an individual's background characteristics and attitudes towards using risk assessment and economic analysis to manage effects of chemical exposures. Contingency table analysis is used to present the bivariate relationships in the data. In addition, the relationships reported are indicated by the degree of association as expressed with summary measures. Interpreting the statistical significance of those relationships (i.e., the likelihood that associations are not due to chance) requires some consideration of the number cases used in the analysis. This is necessary because statistical significance levels are a function of sample size. It generally is easy to obtain statistically significant relationships with large data sets or samples even when the actual degree of association between variables is weak. However, with small samples, it "requires a much more striking relationship in order to obtain significance" (Blalock, 1972: 293). Depending on its strength, to the extent a non-random relationship is established with a limited number of cases, that relationship between variables may be quite important in a substantive sense.

78

Because the variables used in this study involve nominal level measures, both Cramer's V and Goodman and Kruskal's lambda (λ) are employed to indicate the strength of relationships. As a χ^2-based statistic, Cramer's V gives "greatest weight to those columns or rows having the smallest marginals rather than to those with the largest marginals" (Blalock, 1972: 298). The second measure of association, λ, involves an unrestricted minimizing of errors to indicate whether the values of the dependent variable tend to cluster with certain values of the independent variable. Like Cramer's V, lambda is bounded by the range from 0 to 1. However, if one of the row or column marginals is much larger than the rest, λ will assume a zero value and produce a misleading result (Blalock, 1972: 302-303). The values for λ and Cramer's V may be thought of as providing approximate lower and upper bounds respectively of the relationships. Because of this, it also is important to consider the actual frequency distributions of the observed values from which those measures are calculated. Those percentages are presented in the tables along with the measures of association in order to permit such comparisons.

3   FINDINGS

Table 6.1 indicates that most of the employees with training in the social sciences, liberal arts or law are in policy support offices and have policy formulation/implementation responsibilities. Individuals with backgrounds in the environmental sciences are concentrated in the Office of Research and Development (ORD). They primarily perform an information translation function. Individuals whose highest degrees are in the physical sciences or engineering similarly tend to perform an information translation role within EPA. However, they are distributed more uniformly throughout the agency. While one might assume that some statistically significant relationship would exist between education and office of employment or job responsibilities within EPA, the data fail to reveal such an association on a non-random basis.

3.1 Risk Analysis and Benefit-Cost Analysis

Among the regulatory agencies, EPA's institutional mechanisms for risk-benefit analysis may be the most highly developed (see Chapter 4 of this volume; see also U.S. Environmental Protection Agency, 1984; National Research Council, 1983). For example, the EPA has used health risk assessments as a basis for regulating pesticides such as alachlor, captan

Table 6.1 Education, Office and Job Responsibilities[a]

	Education			
	Social Sciences/ Liberal Arts	Environmental Sciences	Physical Sciences/ Engineering	Law
o OFFICE				
Research & Development	11.1	50.0	25.0	0.0
Program	11.1	33.3	37.5	33.3
Policy Support	77.8	16.7	37.5	66.7
Total %	100.0	100.0	100.0	100.0
(N)	(9)	(6)	(16)	(3)

$\lambda = 0.12$ Cramer's V = 0.34 $\chi^2 = 8.063$ $p = 0.23$

	Social Sciences/ Liberal Arts	Environmental Sciences	Physical Sciences/ Engineering	Law
o JOB RESPONSIBILITIES				
Scientific Research	22.2	16.7	18.8	0.0
Information Translation	22.2	83.3	62.5	33.3
Policy Formulation/ Implementation	55.6	0.0	18.8	66.7
Total %	100.0	100.0	100.1	100.0
(N)	(9)	(6)	(16)	(3)

$\lambda = 0.25$ Cramer's V = 0.38 $\chi^2 = 9.686$ $p = 0.13$

[a]Statistical significance probabilities based upon χ^2 values.

or toxaphene under the Federal Insecticide, Fungicide, and Rodenticide Act (U.S. Environmental Protection Agency, 1985, 1986, and 1987). Nonetheless, debate continues as to whether the risk analysis function should be centralized, perhaps within ORD, or remain decentralized among the various program offices. Moreover, EPA continues to wrestle with complex issues regarding duplication of effort, priority-setting and coordination, data collection, and internal review procedures. As a result, the agency initiated a Guidelines Implementation Effort in 1987 in order to make its program and regional offices more aware of the methods and strategies for conducting risk assessments as a basis for regulatory decisionmaking.

Most of the statutes applying to EPA permit an evaluation of costs relative to some reduction in risks, but little guidance as to how much weight benefits should have is offered (Regens et al., 1983). In practice, EPA has utilized cost-effectiveness analysis more frequently because the benefit-cost approach "is often difficult to do, since the Agency is frequently concerned with protecting such things as human life and the stability of ecosystems, social values for which there is no market price, or for which current procedures for finding 'shadow prices' are bitterly controversial" (U.S. Environmental Protection Agency, 1984: 27).

What attitudes toward the use of formal analytical techniques to evaluate risks as well as benefits and costs prevail in EPA? Tabulations of the data from the interviews indicate that an overwhelming majority of the EPA senior personnel interviewed favor the use of formal risk analysis (76.5%). A strong majority similarly support government use of benefit-cost analysis (64.7%).

Do potential cleavages exist among the agency's senior employees in terms of their views about the role of technical analysis in the policymaking process? Such attitudinal differences might stem from divergent epistemological and methodological approaches attributable to varying educational backgrounds. In addition, because the EPA is organized along media as well as functional lines, attitudes may be linked to office of employment and/or job responsibilities. Table 6.2 demonstrates that a strong relationship does exist between education and support for risk assessment. EPA officials with training in the social sciences, physical sciences or engineering are much more supportive of risk assessment than their counterparts. In fact, a strong majority of those with training as environmental scientists oppose its use for policymaking. This may reflect reservations about current capacity to characterize adequately pollutant transport and fate, exposures or effects. Competing explanations for internal divisions receive less support. While ORD and personnel in policy support offices tend to be more supportive, no statistically significant relationship emerged between attitudes toward risk assessment and either office or job responsibilities.

Table 6.2 Support for Risk Assessment[a]

EDUCATION:	Social Sciences/ Liberal Arts	Environmental Sciences	Physical Sciences/ Engineering	Law
Favor	100.0	33.3	81.3	66.7
Oppose	0.0	66.7	18.8	33.3
Total %	100.0	100.0	100.1	100.0
(N)	(9)	(6)	(16)	(3)

$\lambda = 0.25$ Cramer's V = 0.52 $\chi^2 = 9.338$ p = 0.02

OFFICE:	Research & Development	Program	Policy Support
Favor	87.5	50.0	87.5
Oppose	12.5	50.0	12.5
Total %	100.0	100.0	100.0
(N)	(8)	(10)	(16)

$\lambda = 0.0$ Cramer's V = 0.40 $\chi^2 = 5.517$ p = 0.06

JOB RESPONSIBILITIES:	Scientific Research	Information Translation	Policy Formulation/ Implementation
Favor	66.7	77.8	80.0
Oppose	33.0	22.2	20.0
Total %	100.0	100.0	100.0
(N)	(6)	(18)	(10)

$\lambda = 0.0$ Cramer's V = 0.11 $\chi^2 = 0.407$ p = 0.81

[a]Statistical significance probabilities based upon χ^2 values.

Table 6.3 reveals a somewhat comparable pattern in terms of support for benefit-cost analysis among these senior EPA officials. Individuals with environmental sciences backgrounds are strongly opposed while the

Table 6.3 Support for Benefit-Cost Analysis[a]

EDUCATION:	Social Sciences/ Liberal Arts	Environmental Sciences	Physical Sciences/ Engineering	Law
Favor	77.8	33.3	68.8	66.7
Oppose	22.2	66.7	31.3	33.3
Total %	100.0	100.0	100.1	100.0
(N)	(9)	(6)	(16)	(3)

$\lambda = 0.17$ Cramer's V = 0.32 $\chi^2 = 3.379$ $p = 0.33$

OFFICE:	Research and Development	Program	Policy Support
Favor	50.0	50.0	81.3
Oppose	50.0	50.0	18.8
Total %	100.0	100.0	100.1
(N)	(8)	(10)	(16)

$\lambda = 0.0$ Cramer's V = 0.33 $\chi^2 = 3.622$ $p = 0.16$

JOB RESPONSIBILITIES:	Scientific Research	Information Translation	Policy Formulation/ Implementation
Favor	50.0	55.6	90.0
Oppose	50.0	44.4	10.0
Total %	100.0	100.0	100.0
(N)	(6)	(18)	(10)

$\lambda = 0.0$ Cramer's V = 0.34 $\chi^2 = 4.030$ $p = 0.13$

[a]Statistical significance probabilities based upon χ^2 values.

other groups, especially the social scientist group which overwhelmingly consists of economists, are highly supportive. This split may be reflective of an emphasis by the former group on the environment as a valuable entity per se and/or their socialization into a classical public health orientation. Similarly, individuals in policy support offices as well as those with policy formulation/implementation responsibilities tend to be much more favorably oriented towards benefit-cost analysis as a decision tool. However, none of the relationships are statistically significant.

3.2 Distribution of Risk

Three different strategies predominate as options for determining how to distribute risks and benefits. First, individuals may be allowed to make their own choices, with the implicit assumption that they possess adequate information to accept risks and accrue benefits on a voluntary basis. For the second approach, the major decision rule is that aggregate societal costs not exceed aggregate societal benefits. The third approach is grounded in the notion that distributional decisions should be made so that no one group bears a disproportionate share. Such an emphasis on social equity which explicitly takes distribution into account is the method most preferred by our sample of EPA employees. Almost one-half of them opt for that approach (42.4%). The societal net benefits approach is supported by almost one-quarter of the sample (24.2%). In tandem, this suggests strong support for strategies based on concepts underlying microeconomic theory. The net benefits approach received the most support from individuals whose highest degrees were in the social sciences or liberal arts (55.6%) while individuals with physical sciences or engineering backgrounds were most supportive of no group bearing a disproportionate share (60.0%). The social equity method similarly was supported by 50.0 percent of the ORD and policy support office personnel as well as by 58.8 percent of the EPA employees whose job responsibilities involved information translation. It is worth noting, however, that half the individuals with environmental sciences backgrounds were unwilling to endorse any of the three options.

3.3 Valuation of Life

Central to the debate about the utility of benefit-cost analysis is another, possibly more controversial, issue involving whether and how to set a value on human life (see Rhodes, 1980). Among the methods currently being used by economists to place a monetary value on human life, three dominate. The first involves computing lost earnings in the case of premature death or disability and equating this estimate with the value of life/disability. The second method, which is termed contingent valuation, asks individuals how much they would be willing to pay in order to reduce the probability of death or disability due to a given

84

hazard. The third method involves analyzing wage differentials in a series of occupations which involve varying risks of injury and death. The wage differentials are assumed to represent reflections of society's willingness to pay premiums via the labor market for decreases in risk. Quantitative estimates of the economic value of a human life vary depending upon which of the three methods are used to evaluate the monetary benefits of reducing or preventing chemical exposures. The lost earnings technique typically produces the lowest benefit estimates. The contingent valuation approach, on the other hand, generally produces the highest values for benefits (see Bentkover et al., 1986).

EPA's guidelines for performing regulatory impact analyses seem to favor the use of "willingness to pay" measures for benefits estimation. But, the guidelines do permit consideration of alternative approaches, such as the incremental cost-per-life-saved of different regulatory options (U.S. Environmental Protection Agency, 1983: 10-11). This suggests a desire on the part of the EPA to remain flexible on the valuation question, reject a simple decision rule, take into account the wide range of approaches available, and exercise considerable caution in linking these estimates to environmental policy until some very serious methodological and empirical questions can be resolved (see Violette and Chestnut, 1983). In fact, on the general question of whether society must attempt to place an economic value on human life in order to allocate scarce resources, a strong majority of the EPA respondents disagreed (70.6%). Opposition was strongest among individuals with training in the physical sciences or engineering (81.3%), among ORD (75.0%) and program office employees (90.0%), and among those having job responsibilities involving information translation (83.3%).

4   CONCLUSION

Greater understanding of attitudes towards the use of risk assessment and economic analysis has particular relevance for suggested institutional reforms in the way risk management is integrated into an overall approach to environmental protection, especially in designing strategies for managing adverse health effects of chemicals exposures (see Rosenbaum, 1985; Ruckelshaus, 1985; Schmandt, 1984; Regens et al., 1983; Daneke, 1982). Among senior EPA officials, educational background rather than office or job responsibilities appears to play the dominant role in shaping attitudes towards the utility of risk analysis, benefit-cost analysis, methods for the valuation of human life, and distribution of risk. As a result, if the overall performance of the regulatory process is to be enhanced by the use of analytical techniques as an aid for making risk-benefit judgments, these results suggest that efforts to make the environmental management system more effective, efficient, and socially responsible must be based on a sophisticated appreciation of the bases for existing perspectives.

5   ACKNOWLEDGEMENT

This analysis was partially supported by a Research Fellowship from the North Atlantic Treaty Organization, Scientific Affairs Division, Committee on the Challenges of Modern Society. Hans Martin Seip, Anders Heiberg, Karl-Gerhard Hem, Dag Helge Tronnes, and Richard Moraski provided helpful comments. The conclusions, however, are the sole responsibility of the author.

6   APPENDIX

The following provides the specific wording for each of the closed-ended questions used to measure the dependent variables in this study.

1. "Do you favor or oppose the use of formal risk analysis in environmental policymaking?"

2. "Do you favor or oppose the use of benefit-cost analysis in environmental policymaking?"

3. "Scholars differ about the need to consider the <u>distribution</u> for risks and benefits in formulating occupational and environmental regulation. Three dominant perspectives are:

   A. People should be allowed to make their own choices. If a worker decides to work in a risky place, he or she implicitly accepts the risks.

   B. The major rule should be that aggregate societal costs should not exceed aggregate societal benefits.

   C. Decisions should be made so that no one group bears a disproportionate share of the risks.

   Which one of these is closest to your view?"

4. "Three methods are currently being used by economists to place an economic value on human life. I would like to get your opinion on these methods.

   A. Compute the amount of earnings that would be lost in the case of premature death or disability and equate this with the value of life/disability.

B.   Ask individuals how much they would be willing to pay to reduce
the probability of death or disability.

C.   Analyze wage differentials in occupations involving varying
risks of injury and death and use the wage rate differentials
as reflections of societal willingness to pay for decreases in
risks.

If you had to choose a technique for valuing life or injury which
technique would you select?"

7   REFERENCES

Ames, B. N., Magaw, R., and Gold, L. S., 1987, "Ranking Possible Carcino-
       genic Hazards", Science 236:271.

Barker, B., 1984, "Cancer and the Problems of Risk Assessment", EPRI
       Journal, 9:26.

Bentkover, J. D., Covello, V. T., and Mumpower, J., eds., 1986, Benefits
       Assessment, D. Reidel, Boston.

Blalock, H. M., 1972, Social Statistics, 2nd ed., McGraw-Hill, New York.

Daneke, G. A., 1982, "The Future of Environmental Protection: Reflec-
       tions on the Difference Between Planning and Regulating", Public
       Administration Review, 42:227.

Douglas, J., 1985, "Measuring and Managing Environmental Risk", EPRI
       Journal, 10:7.

Epstein, S. S. and Swartz, J. B., 1981, "Fallacies of Lifestyle Cancer
       Theories," Nature, 289:127.

Hoos, I. R., 1983, Systems Analysis in Public Policy: A Critique, 2nd
       ed., University of California Press, Berkeley.

Lave, L. B., 1987, "Health and Safety Risk Analyses: Information for
       Better Decisions", Science, 236:291.

National Research Council, 1983, Risk Assessment in the Federal Govern-
       ment: Managing the Process, National Academy Press, Washington.

Peto, R., 1980, "Distorting the Epidemiology of Cancer: The Need for a
       More Balanced Overview," Nature, 284:297.

Regens, J. L., Dietz, T. M., and Rycroft, R. W., 1983, "Risk Assessment
       in the Policy-Making Process: Environmental Health and Safety
       Protection", Public Administration Review, 43:137.

Rhodes, S. E., ed., 1980, Valuing Life: Public Policy Dilemmas, Westview
       Press, Boulder, CO.

Rosenbaum, W. A., 1985, Environmental Politics and Policy, Congressional
       Quarterly, Washington.

Ruckelshaus, W. D., 1985, "Risk, Science, and Democracy", <u>Issues in Science and Technology</u> 1:19.

Ruckelshaus, W. D., 1983, "Science, Risk and Public Policy", <u>Science</u>, 221:1026.

Russell, M. and Gruber, M., 1987, "Risk Assessment in Environmental Policy-Making", <u>Science</u> 236:286.

Rycroft, R. W., Regens, J. L., and Dietz, T., forthcoming, "Incorporating Risk Assessment and Benefit-Cost Analysis in Environmental Management", <u>Risk Analysis</u>.

Rycroft, R. W., Regens, J. L., and Dietz, T., 1987, "Acquiring and Utilizing Scientific and Technical Information to Identify Environmental Risks", <u>Science, Technology, and Human Values</u>, 12:125.

Schmandt, J., 1984, "Regulation and Science", <u>Science, Technology and Human Values</u>, 9:23.

Swartzman, D., Liroff, R. A. and Croke, K. G., eds., 1982, <u>Cost-Benefit Analysis and Environmental Regulations</u>, Conservation Foundation, Washington.

U.S. Environmental Protection Agency, 1987, <u>Cadmium: Conclusion of Special Review of Intent to Cancel Registrations of Pesticide Products Containing Cadmium</u> OPP-30000/20D, U.S. Environmental Protection Agency, Washington.

U.S. Environmental Protection Agency, 1986, <u>Alachlor Special Review Technical Support Document</u>, U.S. Environmental Protection Agency, Washington.

U.S. Environmental Protection Agency, 1985, <u>Captan Special Review Position Document 2/3</u>, U.S. Environmental Protection Agency, Washington.

U.S. Environmental Protection Agency, 1984, <u>Risk Assessment and Management: Framework for Decision Making</u>, U.S. Environmental Protection Agency, Washington.

U.S. Environmental Protection Agency, 1983, <u>Guidelines for Performing Regulatory Impact Analysis</u>, U.S. Environmental Protection Agency, Washington.

Violette, D. M. and Chestnut, L. G., 1983, <u>Valuing Reductions in Risks: A Review of the Empirical Estimates</u>, U.S. Environmental Protection Agency, Washington.

Wynder, E. L. and Gori, G. B., 1977, "Contribution of the Environment to Cancer Incidence: An Epidemiologic Exercise," <u>Journal of the National Cancer Insititute</u> 58:825.

MODELLING BEHAVIOUR OF POLLUTANTS IN SOIL

FOR RISK ASSESSMENT PURPOSES

C. Lupi

Laboratory of Comparative Toxicology and Ecotoxicology
Istituto Superiore di Sanità
Rome, Italy

1 SOIL DISPERSED CONTAMINANTS: BEHAVIOUR AND RISK FOR HUMAN EXPOSURE

Once a chemical has been released on soil surface (due to accidental leakage or agricultural exposure) it may be subjected to a great number of processes, most of them resulting in the spreading of the pollutant in the environment and in different kinds of human exposure. Human health may be affected by soil pollution via inhalation or ingestion of contaminated dust transported in air, via consumption of water coming from sources sited within contaminated areas, as well as via consumption of vegetables grown on contaminated soils. These ways of exposure, together with the transport processes leading to them, are shown in Figure 7.1. As it can be deduced from

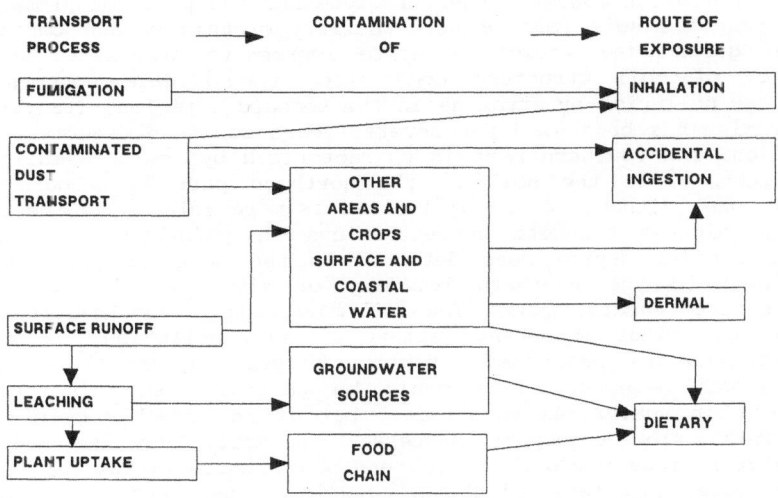

Figure 7.1 Transport of chemicals in soil and human exposure

this figure, the evaluation of risk for humans requires the quantitative evaluation of the rate at which each process will occur. For instance, the evaluation of dietary exposure requires estimation of leaching and plant uptake rates, estimation of the contamination of the food chain and drinking water, as well as the evaluation of the rate of consumption of contaminated food and water. Thus, the evaluation of the overall risk deriving from soil pollution is a very difficult task to deal with, and a screening of the processes most likely to occur in different situations is necessary. In fact, a pollutant may be subjected to very different processes in the same soil under different meteorological conditions. Dust transport and fumigation are more likely to occur in polluted soils sited in arid regions where flow of water in soil is very limited and occasional. In contrast, these two processes are expected to be much less important in rainy regions where flow of water in soil is significant enough to make leaching and surface runoff the most important processes and where frequent rainfall enhances the removal of air dust. Thus, while in arid regions inhalation or accidental ingestion of contaminated dust may be the most common human exposure related to the contamination of soil, it is likely that in rainy regions exposure deriving from consumption of contaminated food or water plays a more important role. Similar considerations can be made taking into account the different partition behaviours of different compounds: substances characterized by a very low solubility in water and high adsorption to organic matter are not considered as potential contaminants of groundwater sources, owing to their negligible mobility in soil.

However, under appropriate meteorological conditions, such compounds may cause considerable risk due to their tendency to be transported in the air as dust-adsorbed contaminants. This is the case of 2,3,7,8-tetrachlorodibenzo-p-dioxin (TCDD), which, in spite of its high toxicity, is not considered as a potential pollutant of drinking water. For this reason, no procedure for the evaluation of groundwater pollution has been included in a recent EPA technical report (Schaum, 1984) which contains a very detailed procedure for the estimation of exposure to contaminated dust.

The third aspect we should consider when evaluating risk deriving from soil pollution concerns the structure and the physical properties of the contaminated soil (mainly permeability, organic carbon content, and depth of groundwater table). A quite impressive demonstration of the importance of soil structure comes from recent data concerning the groundwater pollution by atrazine in the Lombardia region, Italy. In this area atrazine has been used for several years for weed treatment in corn cultivation. The southern part is characterized by low-permeability soil (clay soil), while the soil in the northern part is a carsic, more permeable one. Although corn cultivation is more intense in the southern part, the highest atrazine concentrations in groundwater (measured in wells of similar depth) were detected in the northern part, the peak being close to the northern boundary of the more cultivated, less permeable zone (Funari, 1988). The soil pollution in the Lombardia region also showed another important feature of soil pollution: as mentioned above, atrazine had been used for about 20 years before the groundwater pollution was detected. This could be partially due to the recent improvement in analytical techniques, but it is probably mainly due to the relative slow transport processes in soil. Atrazine release in groundwater sources would not have been significant in the first period, becoming more important with the continued use and the consequent accumulation in soil. These data stress that a time delayed risk for man should be expected following soil contamination. Contamination of groundwater sources may not become evident until many years after the beginning of the pollution of the soil, and may continue also for years after the removal of the cause of pollution.

Risk for human exposure depends not only upon the environmental behaviour of the contaminant, but also upon the characteristics of the exposed population, i.e., habits, age, closeness to the polluted site, and so on. This aspect will not be discussed in detail in this paper, which mainly concerns the environmental behaviour of the pollutants in soil. However, some simple considerations can be made. For instance, children playing in the open air and outdoor workers are likely to be more exposed to dust-contained toxic compounds via inhalation or ingestion of contaminated dust; people living in contaminated areas and eating mainly locally-produced food will be more exposed to contaminated food than people eating food from different regions. This was the case in Seveso where people were mostly eating vegetables cultivated in their own gardens.

These simple considerations illustrate how different the behaviour and hence the health impact of different soil contaminants may be under different patterns of meteorological conditions and soil structure. The quantitative prediction of disappearance from the topsoil, of the rate of release in groundwater, or of the "safe" quantities of pesticides to be applied without risk of contamination of successive crops requires the simultaneous evaluation of too many parameters to be performed without the aid of models. In the last few years, some models were developed for this purpose. Some of these models are based on the numerical solution of the differential equations describing processes occurring in soil, e.g., the model of Leistra et al. (1980), Bromilow and Leistra (1980), and the TOPS and SWIFT model of Lafleur et al. (1981). Other models are based on more empirical approaches which, in general, require a smaller number of parameters. In this chapter the theoretical basis for setting up an "empirical" model of the behaviour of chemicals in soil is discussed. Such a model was then codified and subjected to sensitivity analysis and validation by means of comparison with experimental data reported in the literature.

2 PROCESSES OCCURRING IN SOIL: THEORETICAL BASIS FOR MODELLING

2.1 Adsorption and Partitioning Processes

The soil may be considered as formed by three different phases: air, solid matter (organic and inorganic), and water. The most important property driving the behaviour of pesticides in soil is the way these compounds partition between these three phases. The partitioning process depends upon physico-chemical features of the pollutant and the polluted soil. The adsorption of chemicals to soil solids may be described as following the isotherm law in its linear form:

$$Q = K_d C_{liq}$$

where Q, the amount of solid adsorbed compound, depends upon the adsorption constant K_d and upon C_{liq} the amount of compound present as solute in water. This relationship is mainly valid for reversible adsorption processes. The application of this law to compounds capable of binding irreversibly to soil Organic Carbon (O.C.) may lead to overestimation of the transport of chemicals by means of water flow. However, this overestimation may be considered negligible in the case of the unionized pesticides, which usually bind in a reversible way to the O.C. of soil. The K_d constant depends upon soil adsorption features and upon the physico-chemical features of the involved compound. For a given compound, this constant may vary significantly in different soils; thus, K_d constants should be used for the soil in which they are determined. A better alternative may be the use of the K_{oc} constant (organic carbon/water partition coefficient) which corresponds to the K_d constant normalized for the O.C. content of soil. The use of this constant is based on two widely accepted assumptions (McCall et al., 1983): 1) only the organic carbon fraction of soil is effective for the adsorption processes and 2) the O.C. fractions of different soils have the same adsorption features. The utility of the use of K_{oc} instead of K_d lies in the fact that a single K_{oc} value may be used for modelling purposes regardless of the kind of soil studied. Several relationships exist for the calculation of the K_{oc} value on the basis of few physico-chemical properties of the compound studied (Table 7.1). Thus, the K_{oc} value may be very useful when rapid assessment of the behaviour of pollutants in soil is required and experimental data are not available. However, it should be considered that the use of K_{oc} may be subjected to not easily quantifiable errors due, inter alia, to the effect of clay or other inorganic matters on the adsorption processes and to soil pH. When experimental data are available, the K_d constant may provide more reliable estimates of the partition of the compound between water and soil. The adsorption processes not only influence the rate at which compounds are released to water, but also the rate at which they are released to air. It is, in fact, well known that fumigants are less effective in soils characterized by high O.C. contents. The partition between air, water, and solids may be estimated on the basis of simple compartmentalization models. In Table 7.2 data concerning the distribution behaviour of four different pesticides in two different types of soil are reported.

2.2 Water Flow in Soil

The Darcy Law. A way to describe the vertical water flow in soil is by application of the Darcy law, which can be expressed as:

$$v = -K\frac{\partial H}{\partial L}$$

where v is the flow rate, K is the hydraulic conductivity of soil, L is the vertical distance, and H is the hydraulic head. H may also be expressed as:

$$H = P/\rho g$$

where P is the pressure existing in soil, ρ is the water density, and g is the gravitational constant. Both K and P are related to the saturation state of soil. The parameter P may be calculated assuming a loglinear relationship between P and water content, when water pressure at two or more points of water content in soil is known. Nicholls et al. (1982a,b) assumed soil pressures $= -10^9$ Pa at 0 water content and -1×10^3 Pa at field capacity. The calculation of K is more complex due to the fact that this parameter depends also upon factors such as pore shape and angles, structural composition of soil, etc. Moreover, the fact that the value

Table 7.1 Relationships for the Calculation of K_{OC} (WS = water solubility in ppm; Kow = octanol / water partition coefficient)

$\log K_{OC} = 0.524 \log Kow + 0.381$	(Mill, 1981)
$\log K_{OC} = 3.64 - 0.55 \log WS \pm 1.23$	(Kenaga, 1980)

Table 7.2 Calculated Partition of Four Different Chemicals in Two Different Soils: a) low O.C. content; b) high O.C. content

		% in soil water	% in soil air	% in soil solids	% in atmosphere
Atrazine	a)	5.05	5.24×10^{-7}	94.50	0.45
	b)	1.15	5.19×10^{-8}	98.79	0.06
Carbofuran	a)	21.87	6.66×10^{-6}	72.39	5.74
	b)	6.14	8.10×10^{-7}	92.92	0.94
Fosamine	a)	95.76	1.75×10^{-9}	4.23	1.51×10^{-3}
	b)	83.18	6.60×10^{-10}	16.81	7.63×10^{-4}
TCDD	a)	0.11	8.11×10^{-6}	92.89	6.97
	b)	0.03	8.19×10^{-7}	99.00	0.95

of K at the same water content in soil is very different depending on whether the soil is drying or wetting is well documented in the literature (Marshall, 1967).

Nicholls et al. (1982a) performed a comparison between a model based on the Darcy law and an empirical model. The observed water flow in soil was well simulated by the second model, while the first overestimated the water flow, and required a reduction of the calculated K in order to fit the data. In spite of the difficulties related to the calculation of K, the Darcy law is used in several models for the calculation of both saturated and unsaturated water flow in soil.

Empirical Approach. The difficulties related to the application of the Darcy law may be partially overcome by applying a more empirical approach for the calculation of water flow. Following this approach, we can assume the vertical water flow in soil as being composed of two different types of flow: gravitational and capillary. We assume also that gravitational water flow occurs in soil within the large pores and channels while capillary flow occurs within the capillary channels. With the term "microporosity" we indicate the ratio capillary pores volume / soil volume, while with the term "macroporosity" we indicate the ratio non-capillary pores volume / soil volume. We assume that the microporosity of soil is equal to the field capacity. When the water content of soil exceeds the field capacity (oversaturation state), the major flow in soil is the gravitational flow, which may be considered as a rapid flow not completely in equilibrium with the adsorption processes in soil. In contrast, when the water content of soil is smaller than the field capacity (undersaturation state), the flow in soil is completely driven by the capillary force. An additional assumption is that no flow occurs when the water content of soil corresponds to the wilting point. Dividing the soil into horizontal layers, this set of assumptions may be easily codified in the form of an algorithm (Table 7.3). The required parameters are field capacity and macroporosity of soil. When no field

Table 7.3 Pseudocode for the Empirical Calculation of the Water Flow from the n-th Layer of Soil

---

DO CASE

Gravitational Flow   CASE (Water Content$_n$ > Field Capacity$_n$)

          Water Flow$_n$ = Water Content$_n$ - Field Capacity$_n$

Capillary Flow   CASE (Wilting Point$_n$ < Water Content$_n$ < Field Capacity$_n$)

          Flow$_n$ = Water Content$_n$ - (Water Content$_n$ - Water Content$_{n+1}$)/2

      OTHERWISE   (Flow$_n$ = 0)

      END CASE

Table 7.4 Field Capacity and Pore Volume of Different Soils

Soil type	Total Porosity	Field Capacity
Sand	46%	1.55-5.5%
Sandy Loam	56%	20-22%
Clay Loam	56%	25-30%
Clay	58%	40-46%

Table 7.5 Field Capacity of Soils with Different Clay Content

Average Clay Content (%)	Field Capacity (%)
15	25.9
25	35.4
32	39.8
46	46.6

data are available, these parameters can be estimated on the basis of soil classification (Table 7.4) or on the basis of the average clay content of soil (Table 7.5) (Tombesi, 1977).

2.3 Chemical Transport by Water Flow

Due to the adsorption processes, transport of chemicals by means of water flow may be regarded as similar to liquid chromatography. In fact, adsorption to solids and transport of chemicals occur at the same time in soil, determining the formation of a "peak" of concentration. However, in some respects, transport of chemicals in soil is more complicated than a simple chromatographic process. The first difference lies in the heterogeneity of soil, i.e., different O.C. and water contents and different granulometric features at different depths. In Figures 7.2 and 7.3 the organic content trends of different soils are shown. As can be seen, the O.C. content decreases with depth, and hence the adsorption rate of compound to soil solids also decreases. Another important difference lies in the temporal discontinuity of flow, due to the discontinuity of rainfalls.

Organic Carbon (%)

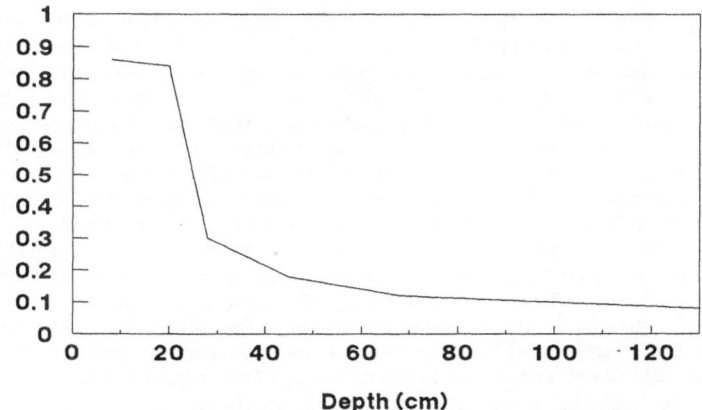

Figure 7.2   Organic carbon trend in a fine-loamy soil

Organic Carbon (%)

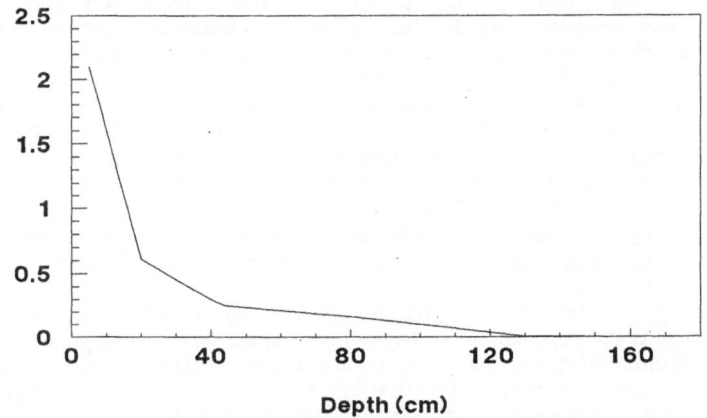

Figure 7.3   Organic carbon trend in a fine-silty soil

The mobility of a compound subjected to a discontinuous flow may be significantly different from that of a compound transported by a continuous, constant flow. Moreover, as mentioned above, the soil is a three-phase system: the overall mobility of compounds in soil is due not only to the convective transport of water flow, but also to the transport due to air flow, which may be important close to the soil surface.

The dynamic interaction between mobilization of the pesticide and adsorption is probably the major difficulty in modelling the behaviour of chemicals in soil. Adsorption and dissolution are not instantaneous processes. Even though the time required by these processes (hours in some cases) is small compared with the time unit (days) used by some models, the assumption of equilibrium between solid and solution may lead to inaccuracy of the predictions, especially in the case of slow adsorption /desorption and rapid water flow. As an example, we consider the case of a compound dispersed on the soil surface and still not adsorbed to soil solids. In the case of rainfall, the compound will be dissolved and then mobilized by water flow before being significantly adsorbed by soil solids. (The equilibrium assumption underestimates the mobility of the compound in this case.) In contrast, a solid-adsorbed compound may be released too slowly by desorption from O.C. in order to be transported by a gravitational water flow at a rate corresponding to the equilibrium between solid and solution. (The equilibrium assumption overestimates the mobility of the compound in this case.) Moreover, in several cases, the existence of more than one peak in the vertical concentration profile of the compound in soil seems to indicate the existence of more than one type of adsorption process.

Some relationships describing the rate of adsorption in soil have been developed in order to avoid the errors due to the equilibrium assumption (for a complete discussion, see Hartley and Graham-Bryce, 1980); unfortunately, none of these relationships have proven completely reliable and the assumption of an instantaneous, reversible adsorption appears to be the only way to describe the mobility of chemicals in soil.

2.4 Plant Uptake

The most common plant uptake processes are considered root uptake, foliar uptake from vapour, and foliar uptake by external contamination of leaves by air transported dust. The uptake by oil cells may be considered a fourth and quite particular process which occurs in some oil-containing plants, and which may play an important role in the uptake of some lipophilic, hydrophobic compounds (Topp et al., 1986). All these processes may represent risk of food-chain contamination when toxic compounds are involved. However, the root uptake represents probably the major process for most organic, low volatile compounds including several pesticides. The application of pesticides to plant roots by means of soil treatment is in fact a widely used practice in crop protection.

The transport system of the higher plant is adapted to carry water and water-solubilized substances. The xylem transport system consists of long vessels connected to form continuous tubes which conduct water and dissolved salts from the roots to the leaves where the water evaporates. This process is commonly called plant transpiration. The first requirement for a compound to be effectively transported by the transpiration stream (which is mainly aqueous) in the xylem is obviously a non-negligible water solubility. However, before entering the roots, a compound must also pass through the endodermic barrier, which is mainly lipidic. Thus, as has been reported in several studies, a compound should also have a moderate lipophilicity in order to enter the transpiration stream (Hartley and Graham-Bryce, 1980; Briggs et al., 1982).

Table 7.6 Daily Mean Rate of Transpiration in g/day

	Birch	Beech	Oak	Larch	Spruce	Pine
Per gram of fresh leaf weight	9.50	4.83	6.02	3.24	1.39	1.88
Mean water consumption in g for the production of one g of dry matter	317	169	344	257	231	300

Table 7.7 Annual Consumptive Use of Water for Different Crops (Kohnke, 1968)

Crop	Annual consumptive use of water (mm)
Alfalfa	690
Grass, hay or pasture	610
Sugar beets	530
Corn	480
Flax	480
Potatoes	480
Small grains	410
Dry beans	360

As far as rate of uptake from soil is concerned, it is important to stress the fact that only the compounds dissolved in soil water can be taken up by roots; thus, the rate of uptake for a compound is expected to depend upon the behaviour of the pollutant in soil, i.e. upon its partitioning between solids, water, and air in soil. Some experimental data reported in the literature confirm this assumption. For instance, Walker (1972) reported that the uptake of pesticides from soil may be considered similar to that from a nutrient culture containing the pesticide at a concentration comparable to those in soil solution as calculated from adsorption behaviour. There is also a general agreement about the fact that the rate of uptake is a passive process not significantly affected by plant metabolism, and that the overall accumulation by plants is linearly correlated with the amount of water transpired (which in turn is a linear function of time).

On the basis of these considerations it appears reasonable to calculate the rate of chemical uptake by a plant simply by multiplying the concentration in soil water by the transpiration rate. The concentration in soil water may be calculated by means of models for the calculation of chemical partitioning in soil. As far as transpiration rate is concerned, estimates reported in the literature (Table 7.6 and 7.7) show that plant uptake requirements vary between 200 to 400 g of water per gram of dry matter produced for the trees, whereas in the case of crops, this requirement may reach the value of 1000 g of water per gram of dry matter (Kohnke, 1968; Donahue, 1985). This indicates that plants may greatly influence water movement, and hence the concentration and mobility of chemicals in soil. Obviously, not all the compounds taken up by plants accumulate in plant tissues; some compounds may in fact be effectively metabolized, or eliminated by means of the renewal of the aerial parts or by means of evaporation-transpiration and photodegradation on the leaves. All these processes are generally dependent upon the different plant species; thus, the rate of biological transformation or elimination of chemicals by plants should be estimated case by case.

2.5 Degradation and Transformation Processes

A great number of degradation and transformation processes occur in soil. A compound is generally subjected to several chemical or biological transformations in soil. Moreover, each process occurs at different rates in the different soil layers. This means that a compound may be subjected to photolysis and/or oxidation in the surface layers of the soil, while the dominant way of degradation in the lower layers may be acid hydrolysis and microbiological transformation. Even though some models for the prediction of the chemical degradation of compounds in soil do exist (Walker and Barnes, 1981), they are highly specific and generally require a large amount of information to be applied. Thus, the use of these model is mainly restricted to comparison with well controlled field or laboratory experiments.

Generally, the relevant soil parameters for the degradation of pesticides in soil are soil moisture, soil temperature, density of microbial fauna, and pH. All these parameters may vary with depth, thus influencing the degradation rate. It is, however, meaningless to include these parameters in a model when no reliable data about their value in the soil and their influence on the degradation rate are known. Very rarely field experiments allow one to estimate these data in natural soils. It is also difficult to evaluate the reliability of assumptions of transformation kinetics different from first order for the overall disappearance of compounds in soil due to chemical transformation (in spite of the fact that some transformation processes following kinetics different from first order in soil may occur in soil under particular conditions). In general, a "safe" assumption is that the degradation rate of compounds released in soil at a rate not exceeding 1-5 kg/ha/year follows first-order kinetics. This assumption is currently adopted by several models (the PRZM model by EPA, and the TOPS model). Nicholls et al. (1982a) reported that the calculated degradation of atrazine and metribuzin using a single first order rate constant was not significantly different either from the degradation calculated by using a first-order rate constant related to moisture content and temperature or from the one calculated by means of the persistence model of Walker and Barnes (1981).

3   AN EMPIRICAL MODEL FOR THE CALCULATION OF MOBILITY OF CHEMICALS IN SOIL

3.1 Model Structure

The model used in this work may be considered as a multi-tank discontinuous model (Figure 7.4). Each layer is represented by three tanks: soil solids with organic carbon, capillary pores and macropores. In order to take into account the different rates at which gravitational and capillary flow proceeds, compounds adsorbed on O.C. are in equilibrium with those in the capillary pores of soil, but not with those contained in the macropores. The rainfall is simulated by adding water to the first layer of soil: if the rainfall exceeds the capillary volume, the water in excess is displaced in the macropores, and then into layer below. The same procedure is adopted for the following layers. In Figures 7.5 and 7.6 algorithms for the calculation of capillary and gravitational flow respectively are shown. The first, modified from Nicholls et al. (1982a,b), is repeated N times for each day of simulation; calculation of upward or downward flow is performed simply by the equilibration of water and solute content in two adjacent layers. This equilibration procedure is repeated until the bottom layer has been reached.

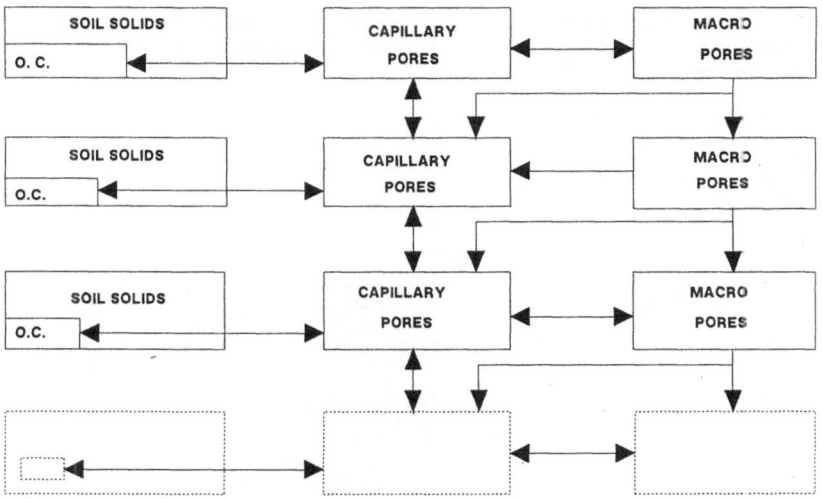

Figure 7.4 Model structure

The second algorithm has been developed for the calculation of gravitational water flow in soil. The first step in this algorithm comprises the search of the first layer (starting from the bottom) in which the field capacity is exceeded. The excess water is then displaced in the total porosity volume of the lower layer. Then the program looks for the excess water in the upper layers until surface has been reached. At this point, the entire water column has moved one layer downward. The procedure is repeated until no water exceeding field capacity exists at any layer in the soil. This may correspond to different situations between two extremes: in the case of low rainfall rate and soil water content, the gravitational flow will stop before reaching the bottom of the soil, when all excess water (if any) has been displaced to the capillary volume. In contrast, in the case of high rainfall rate and water-saturated soil, the gravitational flow will stop when excess water has percolated through the soil and has been released to the groundwater. It follows that the results provided by the model may be significantly influenced by the assumed initial state of saturation of the soil.

As far as the removal by plant uptake processes is considered, the model simply calculates, for each layer and time unit, the quantity of contaminated water entering the roots. The parameters required for the plant uptake process are root depth and averaged transpiration rate. No model for the calculation of plant growth has been adopted. The first-order degradation constant of the compound in soil, obtained from the literature or from experimental data, can be entered as an input for each layer.

3.2 Sensitivity Analysis

A number of simulations were performed in order to obtain information about the relative importance of some parameters, i.e., rainfall rate, rainfall distribution, saturation state of soil, O.C. content of soil, and field capacity value. The "default" values of the parameters subjected to sensitivity analysis with the ranges of variation are reported in Table 7.8. Values for the other model parameters are listed in Table 7.9. The sensitivity analysis was performed by varying only one parameter in each simulation, and setting the others to their "default" values.

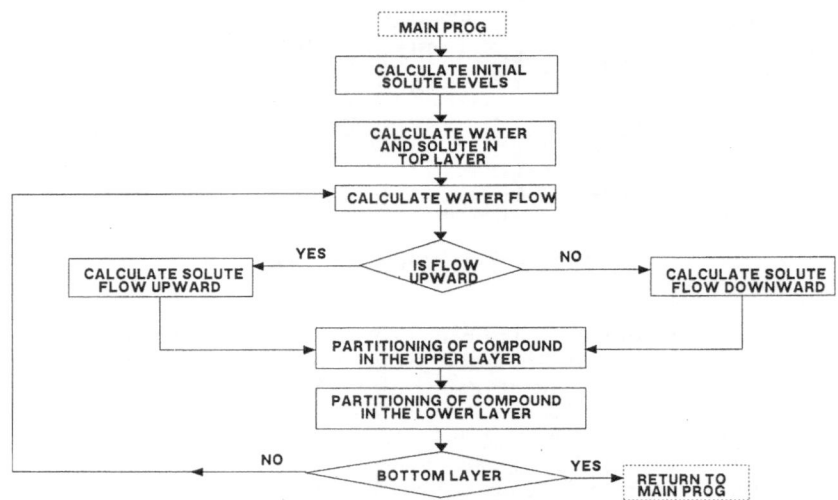

Figure 7.5 Capillary flow flowchart

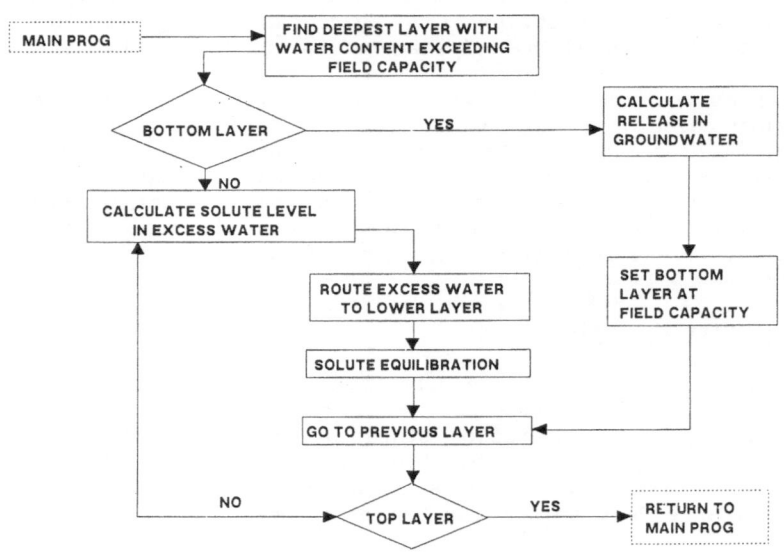

Figure 7.6 Gravitational flow flowchart

Table 7.8 "Default" Values and Variation Range of the Parameters Studied for Sensitivity Analysis

Parameter	"Default" value	Range
Rainfall rate	2 mm/day	1 to 3
Rainfall distribution	4 days	1 to 32
Field capacity	25%	15 to 40
Organic carbon	2%	0.5 to 8
Initial saturation state	50%	15 to 85

Table 7.9 Values of the Parameters not Studied for Sensitivity Analysis

Parameter	Value
Total pore volume	50%
Plant uptake rate	1 mm/day
Plant Uptake Max. Depth	60 cm
Number of layers	15
Layer thickness	5 cm
Number of repetitions of the capillary flow routine	10/day
Pesticide Incorporation Depth	5 cm
Compound:	atrazine
Water solubility	30 ppm
K_{oc}	400
K_w (Reciprocal of the Henry Constant)	8.5×10^6
Half-life in soil	60 days

Figures 7.7 through 7.11 show the results of the sensitivity analysis; all figures were plotted for the 50th day of simulation. In the following paragraphs some details about the sensitivity of the model to each parameter are reported.

Organic Carbon Content (Figure 7.7). A large range of O.C. content was tested. This range covers the values of O.C. normally found in agricultural soils. The model appears, however, to be very sensitive to the value of this parameter, illustrating that it is very important to obtain correct estimates of both the O.C. content of soil and K_{oc} in order to obtain reliable predictions of the behaviour of the compound in soil.

Field Capacity Volume (Figure 7.8). The values for field capacity ranged from 15% (corresponding to a sandy soil field capacity) to 40% (corresponding to the value of a clay soil). As expected, increasing the values for field capacity corresponds to a decrease in the mobility of the compound in soil. The sensitivity of the model to this parameter is moderate.

Initial Saturation State (Figure 7.9). It is interesting to observe the sensitivity of the model for this parameter. A model based on the Darcy law requires the parameter K (hydraulic conductivity of soil) as input. This parameter depends upon the saturation state of the soil, i.e., the larger the saturation state of the soil, the larger the hydraulic conductivity and hence the water flow and the compound mobility. This aspect is simulated well also by the empirical model tested, which does not require any input for the hydraulic conductivity of soil.

Rainfall Rate and Rainfall Distribution (Figures 7.10 and 7.11). As expected, larger rainfall rates result in a larger mobility of the compounds in soil. In contrast, rainfall distribution has only a negligible effect on the amount of leaching at the 50th day.

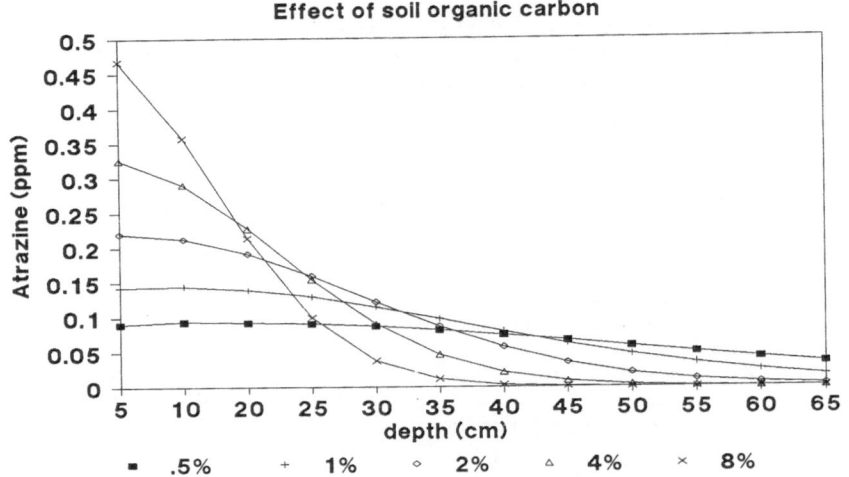

Figure 7.7   Effect of soil organic carbon content

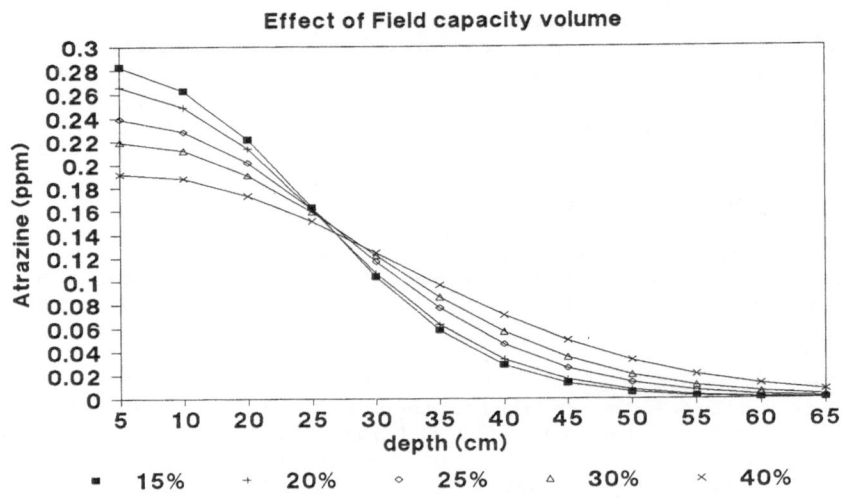

Figure 7.8   Effect of field capacity volume

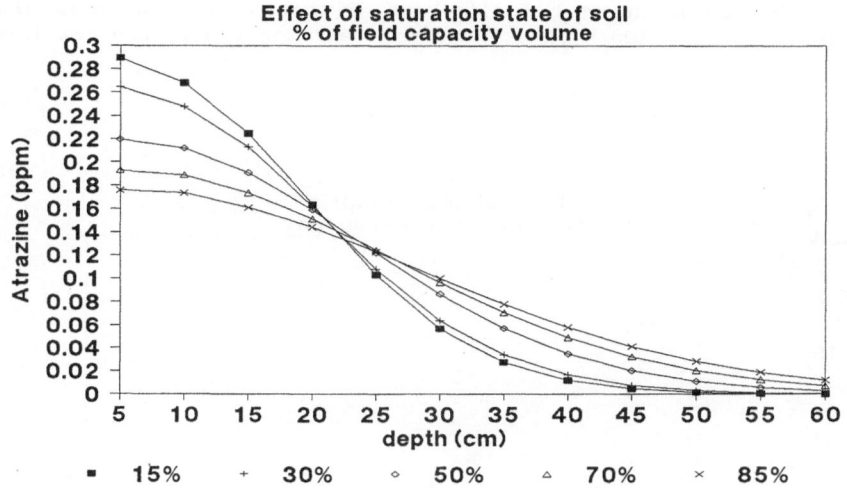

Figure 7.9 Effect of initial saturation state of soil

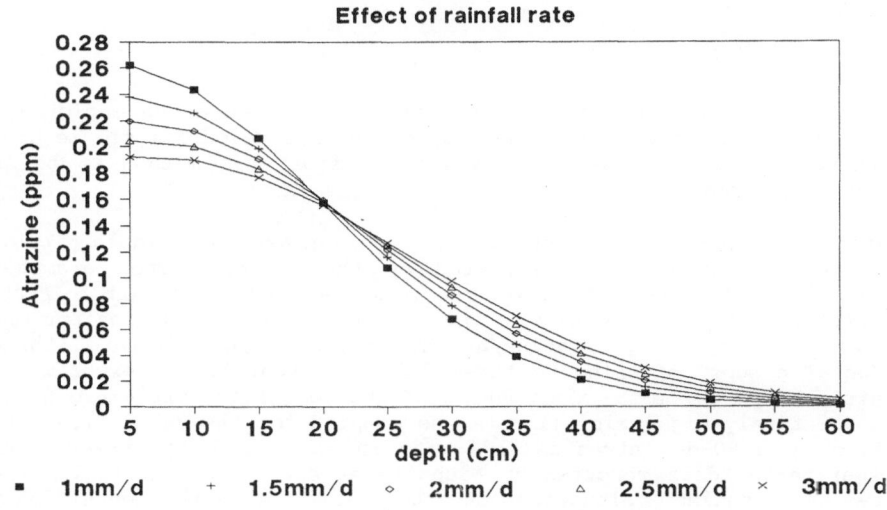

Figure 7.10 Effect of rainfall rate

3.3 Comparison with Experimental Data: Atrazine

Experimental data reported by Basile and Scognamiglio (1983) were used for validation of the model. These authors studied the leaching of atrazine in corn crops treated respectively with 2, 4, and 6 kg atrazine /ha in pre-emergency, using two different irrigation rate corresponding to a total of 200 and 400 m^3/ha. The experiment was replicated for three years.

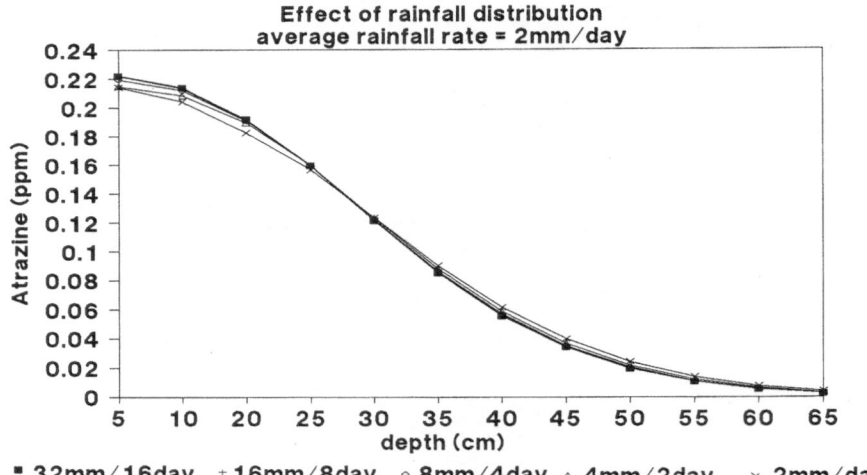

Figure 7.11 Effect of rainfall distribution

The vertical distribution of atrazine was measured at each plot 15, 60, and 120 days after soil treatment by means of a colourimetric method. The authors reported the sensitivity threshold of this method to be in the order of 0.1 ppm.

Data concerning soil structure are given in Table 7.10. Meteorological data were not reported by the authors. The parameter values used for the validation of the model are given in Table 7.11. Figures 7.12, 7.13 and 7.14 provide the results of the validation for one of the experiments (4 kg atrazine/ha, 400 m^3/ha irrigation rate, three experimental groups referring to three different years). The experiments were initiated in the most rainy period of the year (corn pre-emergency); thus, a relatively high rainfall rate was adopted for the validation. The assumption of a 60-day atrazine half-life in soil is in agreement both with experimental data reported by Nicholls et al. (1982a) and with the recovery of atrazine calculated on the basis of experimental data reported by Basile (1981). The field capacity and the macroporosity adopted were extrapolated on the basis of soil composition (Table 7.5). The results of the comparison with the observed data are shown in Table 7.12. The model predicts well the disappearance of atrazine from the first layer in all the cases examined, even though a non-significant tendency to underestimation may be seen for the 120th day. Both the

Table 7.10 Soil Structure (Basile and Scognamiglio, 1983)

Depth (cm)	Coarse sand (%)	Fine sand (%)	Silt (%)	Clay (%)	O.C. (%)
0-20	3.6+0.9	23.3+3.8	32.7+1.4	40.4+4.7	1.54+0.11
20-40	3.6+0.9	24.1+3.2	31.6+2.0	40.7+3.5	1.66+0.11
40-60	3.9+1.1	22.9+4.2	30.9+3.4	42.3+4.7	1.48+0.29

Table 7.11 Parameter Values Used for Validation of the Model

Parameter	Value
Rainfall rate	4 mm/day
Rainfall distribution	15 days
Field capacity	16%
Total porosity volume	40%
Organic carbon	1.5%
Initial saturation state	50%
Plant uptake rate	1 mm/day
Plant uptake max. depth	40 cm
Number of layers	15
Layer thickness	5 cm
Number of repetitions of the capillary flow routine	10/day
Pesticide incorporation depth	0 cm
Compound:	atrazine
Water solubility	30 ppm
K_{OC}	400
K_w	8.5×10^6
Half-life in soil	60 days

calculated and the observed data indicate a higher concentration of atrazine in the first layer than in the lower layers. The model tends to underestimate the concentrations in the bottom layer of soil (40-60 cm depth); in 5 cases of 24 the calculated concentrations are significantly lower than the observed ones. The model also overestimates the concentration of atrazine in the second layer (10-20 cm) at the 15th day. In 42 cases out of 48, the predicted concentrations at the 60th and 120th day are within the confidence limits of the observed data.

These results indicate that the model predicts satisfactorily long-term movement and degradation of atrazine in soil. Problems do exist, however, in the case of short-term predictions. It is very likely that the overestimation of the atrazine in the upper layer at the 15th day is due to a non-instantaneous adsorption to soil O.C., while the model is based on the assumption of instantaneous partition processes. Better predictions could be obtained by including the model in an optimization program, in order to optimize some "uncertain" parameters (e.g. transpiration rate, micro- and macroporosity, initial saturation state of soil, K_{OC}).

4 CONCLUSION

Even though a detailed quantitative description of the different processes occurring in soil may be complex due to the great number of required parameters and to the uncertainty associated with them, some simplifications can be made in order to obtain useful predictions of the behaviour of organic chemicals in soil. The "empirical" models are attractive due to their simplicity and the small number of required parameters. It is, however, important to carefully analyze the "behaviour" of these models in order to verify their reliability in different situations and to obtain reliable estimates of the most sensitive parameters (e.g., K_{OC} and degradation rate). Moreover, due to the generally high variability of field experiments, extensive validation using replicate experiments should be performed.

Table 7.12 Comparison Between Observed and Calculated Concentration in Soil. a = 15th day; b = 60th day; c = 120th day; low = low 95% confidence limit of the observed data; high = high 95% confidence limit of the observed data; calc = calculated by the model. Values in ppm

2 kg atrazine/ha Irrigation rate 200 m³/ha											
	0-10cm			10-20cm			20-40cm			40-60cm		
a)	low	high	calc	low	high	calc	low	high	calc	low	high	calc
b)	0.70	1.10	0.82	0.17	0.20	0.44	0.09	0.12	0.09	0.06	0.12	0.01
c)	0.24	0.58	0.28	0.14	0.34	0.24	0.10	0.20	0.14	0.09	0.12	0.04
	0.12	0.43	0.11	0.09	0.26	0.10	0.02	0.16	0.08	0.09	0.12	0.05

2 kg atrazine/ha Irrigation rate 400 m³/ha											
	0-10cm			10-20cm			20-40cm			40-60cm		
a)	low	high	calc	low	high	calc	low	high	calc	low	high	calc
a)	0.63	0.85	0.81	0.02	0.39	0.44	0.02	0.21	0.09	0.00	0.15	0.01
b)	0.08	0.80	0.27	0.06	0.49	0.24	0.15	0.23	0.15	0.09	0.12	0.05
c)	0.03	0.48	0.10	0.04	0.29	0.10	0.01	0.25	0.08	0.10	0.13	0.05

4 kg atrazine/ha Irrigation rate 200 m³/ha											
	0-10cm			10-20cm			20-40cm			40-60cm		
a)	low	high	calc	low	high	calc	low	high	calc	low	high	calc
a)	1.05	2.48	1.64	0.18	0.32	0.87	0.02	0.29	0.17	0.00	0.31	0.01
b)	0.05	0.89	0.56	0.22	0.30	0.48	0.10	0.30	0.29	0.09	0.12	0.09
c)	0.20	0.41	0.22	0.17	0.23	0.20	0.00	0.32	0.16	0.05	0.20	0.09

4 kg atrazine/ha Irrigation rate 400 m³/ha											
	0-10cm			10-20cm			20-40cm			40-60cm		
a)	low	high	calc	low	high	calc	low	high	calc	low	high	calc
a)	0.76	1.78	1.62	0.19	0.22	0.88	0.05	0.22	0.18	0.01	0.22	0.01
b)	0.14	0.87	0.55	0.06	0.70	0.47	0.13	0.36	0.29	0.06	0.24	0.09
c)	0.08	0.49	0.21	0.15	0.28	0.19	0.01	0.29	0.16	0.03	0.18	0.10

6 kg atrazine/ha Irrigation rate 200 m³/ha											
	0-10cm			10-20cm			20-40cm			40-60cm		
a)	low	high	calc	low	high	calc	low	high	calc	low	high	calc
a)	0.99	2.81	2.45	0.14	0.55	1.31	0.12	0.33	0.26	0.00	0.34	0.01
b)	0.32	2.08	0.85	0.20	0.74	0.72	0.08	0.34	0.43	0.04	0.22	0.13
c)	0.33	0.79	0.33	0.01	0.49	0.30	0.07	0.29	0.24	0.12	0.21	0.14

6 kg atrazine/ha Irrigation rate 400 m³/ha											
	0-10cm			10-20cm			20-40cm			40-60cm		
a)	low	high	calc	low	high	calc	low	high	calc	low	high	calc
a)	2.33	2.80	2.43	0.35	0.55	1.32	0.02	0.38	0.26	0.00	0.43	0.01
b)	0.90	1.10	0.82	0.21	0.86	0.71	0.14	0.45	0.44	0.07	0.19	0.14
c)	0.30	0.82	0.31	0.24	0.53	0.29	0.04	0.31	0.23	0.01	0.22	0.14

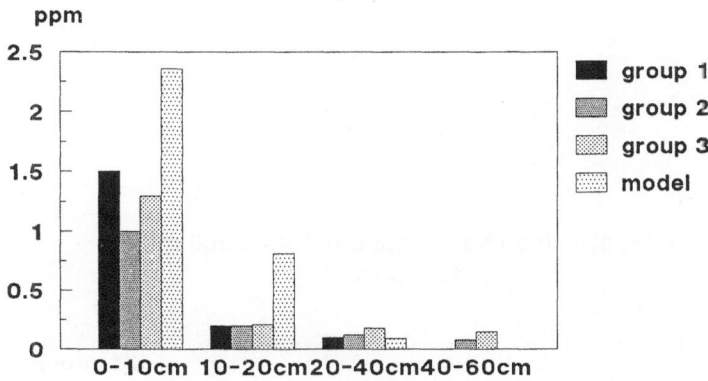

Figure 7.12 Comparison between observed and calculated data: 15th day

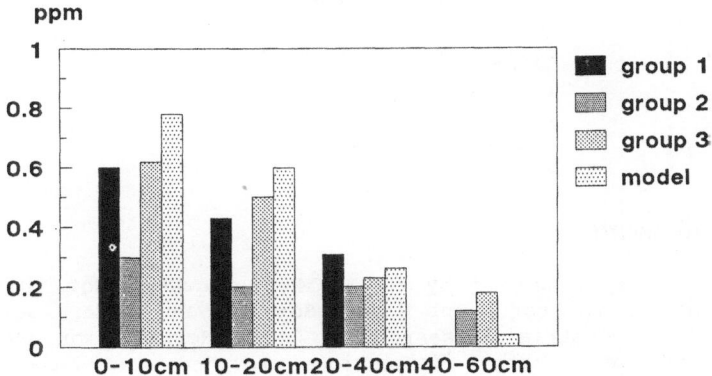

Figure 7.13 Comparison between observed and calculated data: 60th day

It is probable that the major source of error in these kinds of models lies in the assumption of instantaneous adsorption; the possibility of errors due to this assumption should be considered when using these models for obtaining estimates concerning the risk for man from soil contamination. Unfortunately, very little is known about the kinetics of the adsorption processes in soil; thus, the reliability of assumptions different from the one of instantaneous equilibrium may be difficult to evaluate.

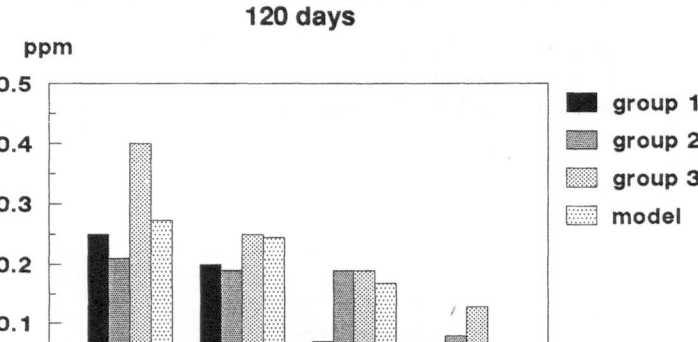

Figure 7.14 Comparison between observed and calculated data: 120th day

5   ACKNOWLEDGEMENTS

This work was supported by NATO-CCMS fellowship No. 404/86. The author is grateful to Prof. Hans Martin Seip and Dr. Anders Heiberg, of the Center for Industrial Research of Oslo, Norway, for their kind hospitality and the important suggestions provided. The author is also grateful to the US Environmental Protection Agency, and particularly to Dr. J. Falco for having made possible the visit to the EPA laboratories, to Dr. R. Moraski for the suggestions and the material provided, and to the entire staff of the Environmental Research Laboratory of Athens, Georgia for having patiently illustrated the activities of their laboratory and for having furnished precious material concerning water quality modelling.

6   REFERENCES

Basile, G., 1981, Interazione fra i minerali argillosi e la sostanza organica con gli erbicidi, in: "Inquinamento del terreno: la mobilità degli erbicidi nel suolo in relazione all'inquinamento delle acque di falda", Consiglio Nazionale delle Ricerche, collana del progetto finalizzato "Promozione della qualità dell'ambiente", AC/C/165-174.

Basile, G., and Scognamiglio, D., 1983, Mobilità e degradazione dei residui di Atrazina nel suolo, Inquinamento, 7/8:39.

Briggs, G. G., Bromilow, R. H., and Evans, A. A., 1982, Relationships between lipophilicity and root uptake and translocation of non ionized chemicals by barley, Pestic. Sci.,13:495.

Bromilow, R. H., and Leistra, M., 1980, Measured and simulated behaviour of Aldicarb and its oxidation products in fallow soils, Pestic. Sci., 11:389.

Donahue, R. L., 1985, "Soils. An Introduction to Soils and Plant Growth", second edition, Prentice-Hall, Inc., Englewood Cliffs, N.J.

Funari, E., Brambilla, A.L., Camoni, I., Canuti, A., Cavallaro, A., Chierici, S., Cialella, G., Donati, G., Jaforte, A., Prandi, L., Salamana, M., Silano, V., and Zapponi, G. A., 1988, Atrazine pollution of drinking water in Italy, Ecotox. and Environ. Safety, (in press).

Hartley, G. S., and Graham-Bryce, I. J., 1980, "Physical principles of pesticide behaviour", Academic Press, London.

Kenaga, E. E., 1980, Predicted Bioconcentration factors and soil sorption coefficient of pesticides and other chemicals, Ecotox. and Environ. Safety, 4:26.

Kohnke, H., 1968, "Soil Physics", McGraw-Hill Book Company, New York.

LaFleur, D. W., Pearson, F. J., and Ward, D. S., 1981, Mathematical Simulation of Aldicarb Behaviour on Long Island: Unsaturated Flow and Ground Water Transport, Report by the INTERA environmental consultants, Inc. for the U.S. Environmental Protection Agency.

Leistra, M., Bromilow, R. H., and Boesten, J. J. T. I., 1980, Measured and simulated behaviour of Oxamyl in fallow soils, Pestic. Sci., 11:379.

Marshall, T. J., 1967, Relations Between Water and Soils, Technical Communication No. 50, Commonwealth Bureau of Soils, Harpenden.

McCall, P. J., Laskowski, I. A., Swann, R. L., and Dishburger, H. J., 1983, Estimation of environmental partitioning of organic chemicals in model ecosystems, Residue Reviews, 13:231.

Mill, T., 1981, "Minimum data needed to estimate environmental fate and effects for hazard classification of synthetic chemicals", Proc. Workshop on the Control of Existing Chemicals (OECD), June 10th, 1981, West Berlin.

Nicholls, P. H., Walker, H., and Baker, R. J., 1982a, Measurement and simulation of the movement and degradation of atrazine and metribuzin in a fallow soil, Pestic. Sci., 12:484.

Nicholls, P. H., Bromilow, R. H., and Addiscott, T. M., 1982b, Measured and simulated behaviour of fluometuron, aldoxycarb and chloride ion in a fallow structured soil, Pestic. Sci., 13:475.

Schaum, J., 1984, "Risk analysis of TCDD contaminated soil", U.S. Environmental Protection Agency, Office of Health and Environmental Assessment, Office of Research and Development, EPA-600/8-84-031.

Tombesi, L., 1977, "Elementi di Scienza del Suolo e di Biologia Vegetale", Edagricole, Bologna.

Topp, E., Scheunert, I., Attar, A., and Korte, F., 1986, Factors Affecting the Uptake of ^{14}C-labeled Organic Chemicals by Plants from Soil, Ecotox. and Environ. Safety, 11:219.

Walker, A., 1972, Availability of Atrazine to Plants in Different Soils, <u>Pestic. Sci.</u>, 3:139.

Walker, A., and Barnes, A.,1981, Simulation of a herbicide persistence in soil; a revised computer model, <u>Pestic. Sci.</u>,12:123.

ENVIRONMENTAL AND HEALTH IMPACT ASSESSMENT OF SOIL POLLUTANTS:

THE SEVESO ACCIDENT AS A TYPICAL EXAMPLE

G. A. Zapponi and C. Lupi

Istituto Superiore di Sanità
Rome, Italy

1   INTRODUCTION

It is well known that the health impact of soil-contained contaminants may be both direct and indirect and can be exerted through complex environmental pathways and food-chains.

In this paper, some aspects of health and environmental risk assessment of environmentally persistent soil contaminants are discussed. The environmental behaviour of the 2,3,7,8-tetrachlorodibenzo-p-dioxin (TCDD) is examined in detail. Due to its toxicological and physical-chemical properties, this compound could represent a major risk to human health, if present in the environment. A large number of experimental data, mostly produced after the Seveso accident, is available to identify and assess its environmental behaviour. This compound can be considered a typical example of an environmentally persistent, bio-accumulating, highly toxic and potentially carcinogenic pollutant.

From a general point of view, the health impact of soil-contained contaminants is mainly exerted through the following pathways:

- Direct Soil-Man transfer: through ingestion, inhalation and dermal absorption of dust and particles originating from contaminated soil.

- Soil-Plant-Man transfer: through plant contamination (due to plant absorption of soil contaminants and to plant contamination by soil originating dust) and subsequent human alimentary use of contaminated plant tissues.

- Soil-Animal-Man transfer: through animal contamination (due to ingestion, inhalation and dermal absorption by animals of dust and particles originating from contaminated soil) and subsequent human alimentary use of animal contaminated products.

- Soil-Plant-Animal-Man transfer: through ingestion by animals of contaminated plants and subsequent human alimentary use of animal contaminated products.

- A combination of the above described pathways (Poc-chiari et al., 1986)

The detected and estimated environmental behaviour of TCDD, capable of causing these processes, will be briefly described here.

2   THE SEVESO ACCIDENT

On July 10, 1976, an uncontrollable exothermic process took place in a reaction bulk during the synthesis of trichlorophenol at the Givaudan-La Roche ICMESA plant at Seveso, 30 km north of Milan, Italy. The temperature inside the reactor rose far above 200 C (probably, up to 400 C)$_2$ and the pressure reached the safety valve critical point (about 3 kg/cm^2) causing the release of a toxic cloud in the open atmosphere. At the moment of the environmental release, the principal volatile compound inside the reactor was ethylene glycol. It has been estimated that nearly 3000 kg of organic matter were released from the reactor, including at least 600 kg of sodium trichlorophenate with some amounts of 2,3,7,8-TCDD (Inquiry, 1979; Di Domenico et al., 1982a; Pocchiari et al., 1983). The visible part of the cloud rose up to about 50 m, and then was driven approximately southeast by wind. Nearly 1800 ha of a densely populated area, called the "Brianza of Seveso" (Figure 8.1), were contaminated. Close to the ICMESA plant, leaves of plants, courtyard animals and birds were seriously affected, many dying within a

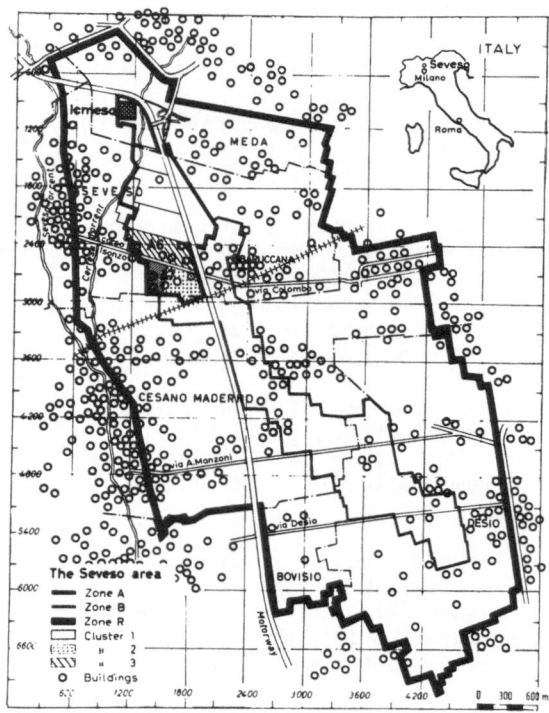

Figure 8.1   The TCDD contaminated area after the Seveso accident

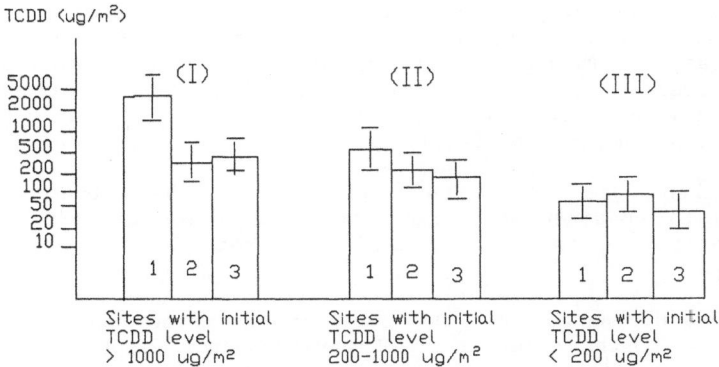

Figure 8.2   TCDD time trends in soil from sites at different contamination levels (I, II, III). 1: average level at 1 month; 2: average level at 5 months; 3: average level at 17 months (after the environmental dispersion)

few days after the accident. At the same time, dermal lesions began to appear among the inhabitants of the area.

Only 9 days after the accident it was assessed that 2,3,7,8-TCDD was present in various types of environmental samples collected near the ICMESA plant. On July 26, 1976, Italian authorities evacuated 179 people from a 15 ha area immediately southeast of the plant. Soon after, further sampling of soil and vegetation indicated the presence of TCDD in a more extended area. As a consequence, all the inhabitants (733 people) in a wide area (approximately 110 ha), coded "Zone A", were evacuated. Moreover, inhabitants of the surroundings (Zones B and R) were subjected to a number of hygiene regulations, including the prohibition to farm and consume local agricultural products and keep poultry and other court-yard animals. Zone B (about 270 ha) was the natural southeastwards extension of Zone A and exhibited lower TCDD levels (Figure 1). Zone R (1430 ha), exhibiting near detection or undetectable TCDD levels, enclosed both Zone A and B. The Zone A-Zone B_2 and the Zone B-Zone R borderlines ran approximately along the 50 ug/m^2 and 5 ug/m^2 mean concentration lines respectively. The Zone R external boundaries were set at undetectable TCDD level (formally, < 0.75 ug/m^2 at that period).

The emergency phase lasted up to the end of the first week of August. After this period, a more precise and detailed survey of the extent and consequences of TCDD environmental distribution was planned and initiated, together with an assessment of the TCDD environmental fate and persistence.

3   ENVIRONMENTAL PERSISTENCE OF TCDD

The environmental persistence of TCDD in the Seveso contaminated area has been already discussed in several works (Di Domenico et al., 1980; Pocchiari et al., 1983). The average time trend of TCDD soil levels, measured at the same sampling points at different time periods, was characterized by a significant initial decrease and by a much more limited subsequent reduction or even a substantial absence of significant variations. This finding may be explained by the detected translocation process of TCDD from the soil surface to deeper soil layers and, thus, by the consequent progressive decrease of the solar ultra violet (UV) degradation process efficiency.

In this work, the above mentioned data have been re-examined separately, considering sites at high, medium and low TCDD concentrations in the most contaminated area (Zone A) of the Seveso territory.

Figure 8.2 shows the TCDD levels in the three groups of sites, at 1, 5 and 17 months after the TCDD environmental dispersion. The three groups include sites whose TCDD level at approximately one month after the accident was respectively: i. higher than 1000 ug/m^2, ii. between 200 and 1000 ug/m^2, and iii. lower than 200 ug/m^2. The time trends were significantly different for these three groups of sites (p <0.01). In particular, at highly contaminated sites the TCDD levels appeared to decrease about 10-fold in the period from the first to the fifth month. At medium-level contaminated sites the TCDD levels decreased about 2-fold in the same time period, while they remained practically constant at low-level contaminated sites. Figure 8.3 shows the three estimated time trends. In all cases, the TCDD level decrease was very limited or practically non-existent after the fifth month.

4   VERTICAL DISTRIBUTION OF TCDD IN SOIL

The vertical distribution of TCDD in the Seveso contaminated area has been widely discussed in previous papers (Di Domenico et al., 1980;

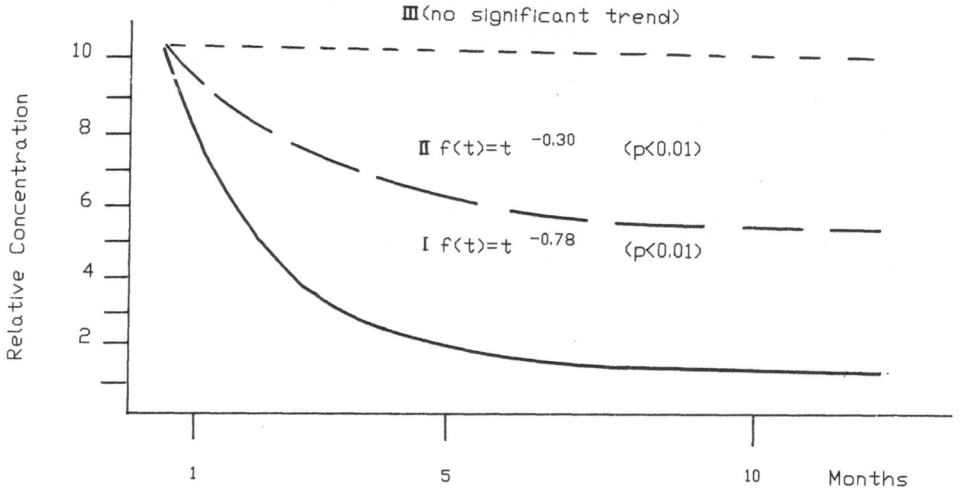

Figure 8.3   TCDD estimated time trends in soil

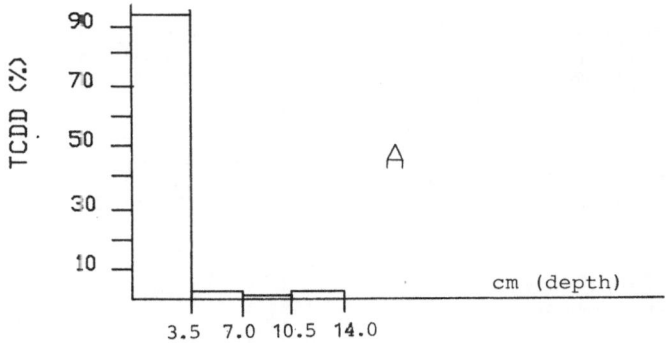

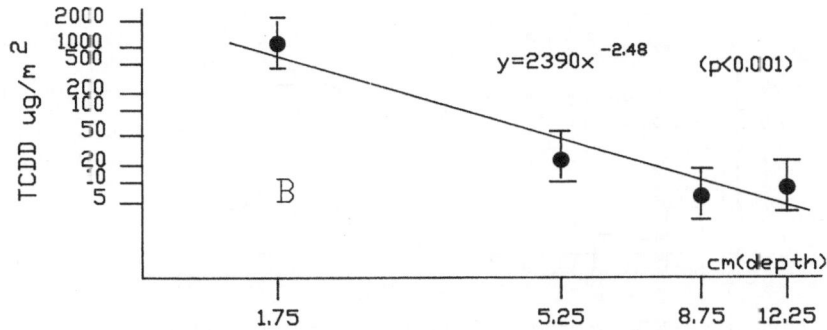

Figure 8.4 TCDD vertical distribution in soil 1.5 months after
contamination. Average of data from sampling sites A_1, A_2
and A_3 (Fig. 8.4 A) in the most contaminated area, and
corresponding linear regression with logarithmic axis (Fig.
8.4 B)

Pocchiari et al., 1983). Over 90% of the detectable TCDD was generally
found in the upper 15 cm thick soil layer. Results of samplings relative
to the 1976-1977 period indicated only a slight change in vertical
trends. In this work the TCDD vertical distribution features at sites
with different TCDD contamination levels have been examined.

Figure 8.4 shows the average TCDD vertical distribution at three
sites within the most contaminated area, detected 1.5 months after the
environmental release of TCDD.

Figure 8.5 shows the TCDD vertical distribution detected at two
sites sampled at different time periods. These data indicate that most
TCDD permeation in soil took place shortly after the accident. Five
months after the contamination of the soil, very limited changes were
detected in the TCDD soil vertical distribution. Figure 8.6 shows the
trend of the ratio of the TCDD contents in the 0-7 cm and in the 7-14
cm deep soil layers. These data, produced in a systematic survey of TCDD
vertical mobility and distribution (Di Domenico et al., 1980), suggest
that, at about one month after the accident, the percentage of the
soil-contained TCDD which permeated below the first 7 cm thick soil
layer was generally lower in highly contaminated sites than in slightly
contaminated sites.

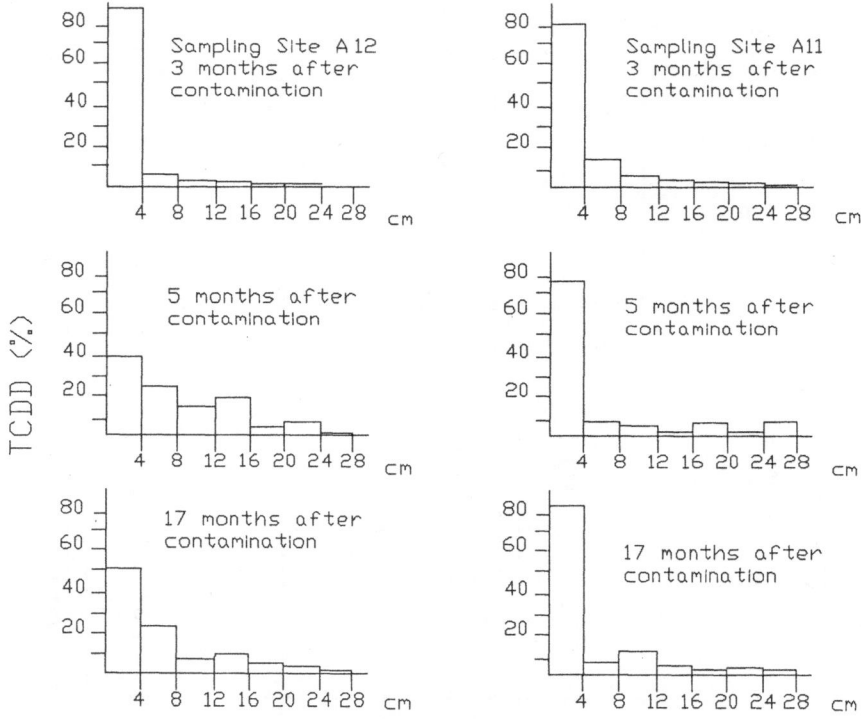

Figure 8.5   TCDD vertical distribution in soil at different time
periods

Figure 8.7 shows the TCDD vertical distribution in the very top
soil layer (0-2 cm deep), detected in northern Zone A, close to the
sampling sites whose assessed vertical distribution is shown in Figures
8.4 and 8.5 (in Figure 8.7, the distribution is expressed as percentage
of the whole TCDD amount in the 0-2 cm deep soil layer).   The
distribution determined in the very top soil layer was not in agreement
with the trend extrapolated from the lower layer values (indicated by
the dashed line in Figure 8.7).

In particular, the TCDD levels detected on the soil surface were
much lower than expected from such a trend. The ultra violet photodegradation process which can be expected to have taken place on the soil
surface is the most reasonable explanation for this finding.

5   PLANT CONTAMINATION AND SOIL-TO-PLANT TRANSFER OF TCDD

The Lombardia Region and the Provincial Chemical Laboratory of
Milan carried out several systematic monitorings of TCDD concentrations
in plants growing naturally or cultivated in experimental fields in
selected areas of the Seveso contaminated territory (Lombardia Region

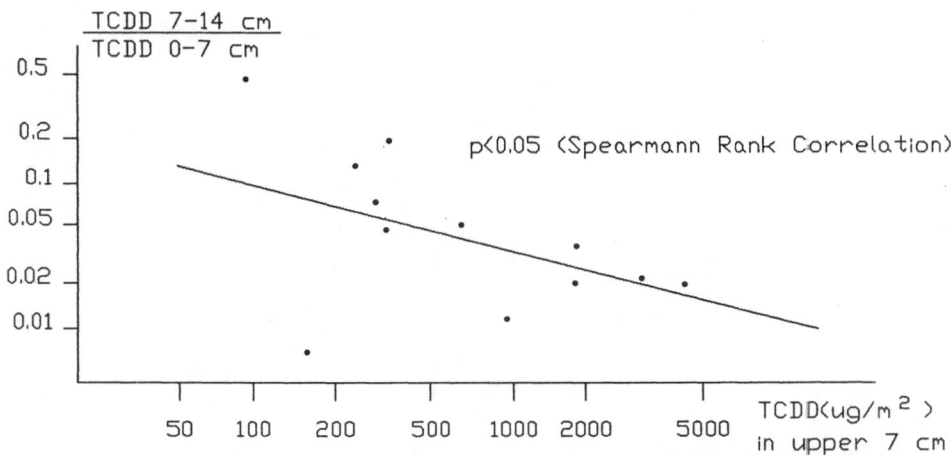

Figure 8.6   Trend of ratio of TCDD concentration in the 0-7 cm and 7-14
cm deep soil layers - August 1976 data

and Provincial Chemical Laboratory of Milan reports, 1976 - 1980).

The statistical distribution of TCDD concentrations detected in plants from the Seveso contaminated area was typically "Log-Normal", as was already established for TCDD concentrations in soil. Such a specific distribution was found also in areas of limited surface and homogeneous soil characteristics, with a relatively limited TCDD level variability. This finding may be reasonably interpreted as the consequence of the original contamination process, which seems to have taken place mainly in the form of "discrete" particles settling on the surface of the affected soil. This was seen from the direct observation of the damages caused to affected leaves. As is well known, the number of drops (e.g., rain drops) or particles falling from a continuous source on a given surface per time unit is described by the Poisson statistical distribution. This kind of deposition process, together with the expected variability of dimensions and TCDD content of the contaminated particles, may well explain the assessed statistical distribution of TCDD levels in soil, which is asymmetrical and tailing off towards higher TCDD values (analogously to a Poisson distribution). The logarithmic transformation of TCDD soil levels was found to produce normality of distribution (Log-Normal distribution of original data) (Di Domenico et al., 1982,b).

The statistical distribution of TCDD concentrations in plants could be explained as the consequence of the same type of contaminating process, as well as of the soil-to-plant TCDD transfer process.

Figure 8.8 shows the statistical distribution of TCDD concentrations detected in epigeal parts of fodder sampled in the Seveso Zone A fields. The average TCDD levels in the fields' soil were in the order of 200-400 ug/m^2 (about 2-3 ppb). The geometric mean of TCDD levels in this fodder was about 0.02 ppb, with a 95% confidence interval of 0.01 - 0.03 ppb; the standard deviation of the log-transformed data was about 1.3, corresponding to a factor of about 3.5 for the original

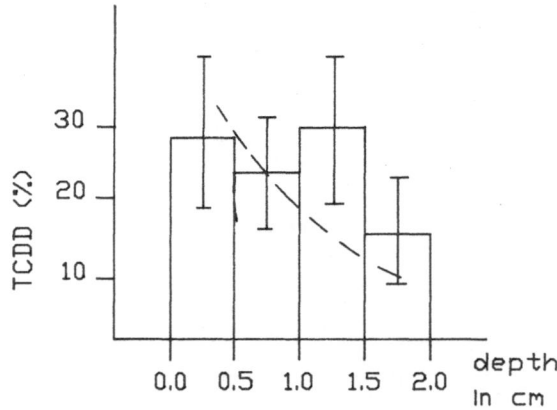

Figure 8.7   Vertical distribution of TCDD in the very top soil layer
(0-2 cm, Zone A). The dashed line represents the distribution extrapolated from deeper soil layer values

data (thus indicating a very high variability). The TCDD variability in
fodder was about two times the variability in soil (as described by the
log-transformed data standard deviations and the corresponding factors
for the original data).

Figure 8.9 shows the statistical distribution of the ratio between
the TCDD concentrations detected in hypogeal and epigeal parts of the
examined fodder. The geometric mean of this ratio was about 6, with a
95% confidence interval of 3 - 11. The geometric mean of the ratio of
TCDD concentration in fodder epigeal part and the concentration in the
corresponding soil was about 0.03, with a 95% confidence interval of
about 0.02 - 0.06. The same parameter, in the case of a sample of
bulbous plant hypogeal parts, was about 0.6, with a 95% confidence
interval of about 0.4 - 0.9.

As a rule, a systematically higher TCDD content was found in
hypogeal parts rather than in epigeal parts of plants, the concentration
levels being close to corresponding soil levels in the case of bulbous
plant hypogeal parts. This result is only to a minor extent ascribable
simply to a direct surface contamination by soil dust of hypogeal parts
of plants. In fact, plant tissues were carefully washed before chemical
analysis and generally only internal parts of bulbs were examined.

In order to better investigate TCDD plant contamination processes,
experiments were carried out outside the Seveso area, using plants
cultivated in greenhouses under controlled conditions (Facchetti et al.,
1984; 1986). These experiments showed that the ratio between the TCDD
concentrations in plants and in the corresponding soil was not constant.
The ratio appeared to change both with time and with the level of TCDD
in soil. This confirmed, at least in part, results obtained in the field
(Cocucci et al., 1979; Cocucci, 1980; Pocchiari et al., 1983).

Fig. 8.10 shows the regression of TCDD concentration detected in
the maize and bean roots and in the corresponding soil contained in the

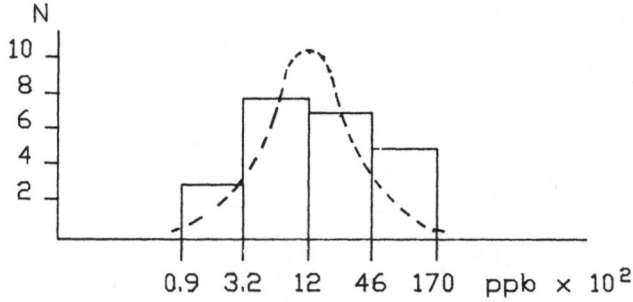

Figure 3.8   Statistical distribution of the TCDD concentration in
the epigeal part of fodder sampled in Seveso Zone A

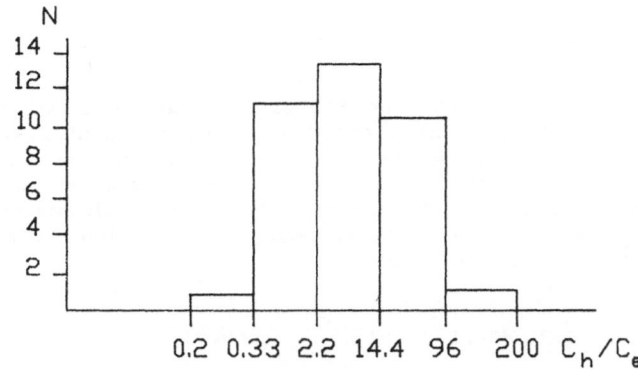

Figure 8.9   Statistical distribution of the ratio of TCDD concentrations
in hypogeal and epigeal parts of fodder from Seveso Zone A

cultivation pots used in the greenhouse experiments. The estimated
function is:

$$(TCDD_{roots}) = 16.3 \ (TCDD_{soil})^{0.612} \quad (p<0.01)$$

$$(TCDD \ level \ in \ ppt)$$

Plant roots were washed before the chemical determination of their TCDD
content. The sediments present in the water used to wash the plant roots

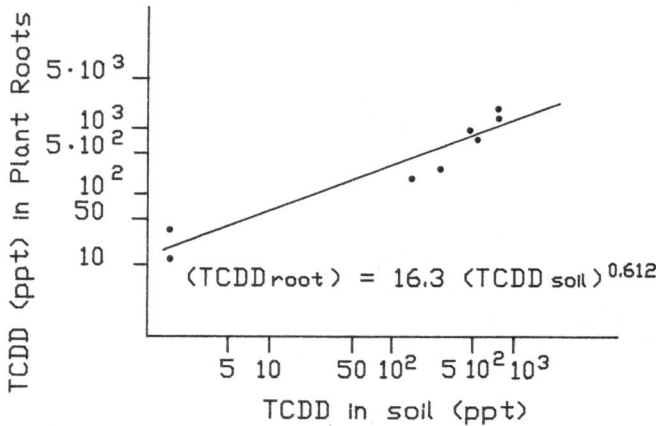

Figure 8.10   Regression of TCDD concentrations detected in maize and
bean roots and in the soil of the corresponding cultivation
pots

were collected and analyzed. These sediments are assumed here to be
representative of soil particles and dust in direct contact with roots.
Figure 8.11 shows the regression of TCDD concentration in the above
mentioned roots and in the sediments from root washing, while Figure
8.12 shows the regression of TCDD concentration in such sediments and in
the soil contained in the corresponding cultivation pots. The two
estimated functions are respectively:

$$(TCDD_{roots}) = 1.36 \ (TCDD_{sed.})^{0.971} \quad (p<0.01)$$

$$(TCDD_{sed.}) = 30.9 \ (TCDD_{soil})^{0.576} \quad (p<0.01)$$

$$(TCDD \ level \ in \ ppt)$$

These data indicate comparable TCDD concentrations in roots and in the
soil particles in direct contact with them . The relationship between
the TCDD concentration in plant roots and the average TCDD concentration
in the soil in the pots where the plants were cultivated appeared to be
non linear. More precisely, in the range of data considered, the
regression function shown in Figure 8.10 indicates a TCDD concentration
in plant hypogeal tissues approximately proportional to the square root
of the concentration in the corresponding soil (it is important to
notice, however, that this regression function is significantly
influenced by the data corresponding to the lowest TCDD levels both in
plant and in soil). The above reported function, if applied to the data
obtained in the experimental fields of Seveso Zone A (TCDD soil average
concentration in the order of 4000 ppt), indicates a value of about 0.6
for the ratio "TCDD in plant/TCDD in soil", consistent with the
measurements obtained in the case of bulbous plant hypogeal tissues.
Further experiments carried out in the field (Seveso Zone A and B)

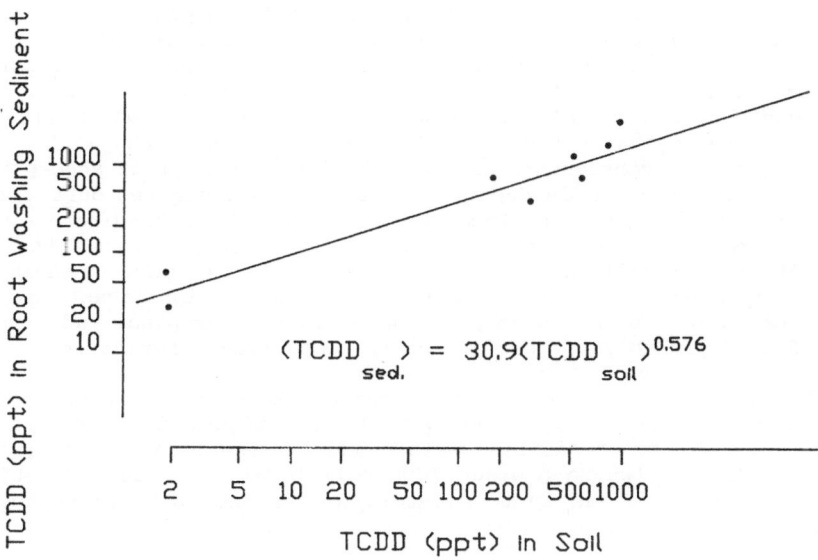

$$(TCDD_{sed.}) = 30.9(TCDD_{soil})^{0.576}$$

Figure 8.11   Regression of TCDD concentrations detected in the sediments from root washing and in the soil of the corresponding cultivation pots

generally indicated a higher ratio "TCDD in plant/TCDD in soil" at low-level contaminated sites than the one expected on the basis of highly contaminated soil data (Pocchiari et al., 1986).

The very low water solubility of TCDD may offer a key to explain these results. TCDD solubility has been reported as approximately equal to 0.2 ppb (Esposito et al., 1980; Hay, 1982); a recent study indicates a much lower value, equal to about 8 ng/l or 8 ppt (Adams and Blaine, 1986). The TCDD water solubility obviously represents the upper limit for the transfer from soil to soil-contained water. When this limit is reached, the partition constant between soil-contained water and soil-contained organic carbon may not be expected to further govern the ratio of TCDD concentrations in these two substrata. Under such conditions, the TCDD concentration in soil-contained water will remain constant (at a level equal to its solubility) even if its concentration in soil solids increases indefinitely. This also implies an upper limit for TCDD absorption, with water by plant roots, in highly contaminated sites.

The role of a significant water transport of TCDD from soil to plant roots seems to be confirmed by the experiments of Facchetti et al. (1984; 1986), which indicate a much higher TCDD concentration in the soil particles in contact with roots (plant root washing sediments) than in the surrounding soil. The water flow from soil to plant roots may range from 10 to 400 liters or more per year for the plants considered here (depending on soil characteristics, plant species, rainfall level, etc.). Even if TCDD solubility is extremely low, this flow may be enough to cause a significant plant contamination.

In addition, other processes may contribute to plant contamination. First, the transport by water of soil particles containing organic carbon (to which TCDD can be adsorbed) may highly increase TCDD mobility with water, particularly in the first 10-20 cm deep soil layer, where the organic carbon is principally present. This process may significantly contribute to the increase in TCDD concentration in the soil close to plant roots as a consequence of water flow caused by root absorption of water. Second, the simple deposition of TCDD contaminated dust originating from soil on plant surface and a possible TCDD binding to it may also be important. Facchetti et al. (1984; 1986) suggest that TCDD evaporation from soil may have an important role in plant contamination, especially under laboratory conditions without a significant air turnover. The whole process is complex and a single conceptual and mathematical model is probably not sufficient to completely describe it.

On the basis of the above discussed data and criteria, it was decided that comprehensive analytical investigations were essential before removing the prohibition to farm and consume agricultural products in Seveso Zone B (whose average TCDD soil level was about 12 ppt, with values in limited areas up to 2.5-fold higher and sporadic values even higher) (Pocchiari et al., 1986).

6   SOIL-TO-AIR TRANSFER OF TCDD CONTAMINATED DUST

A continuous monitoring of atmospheric dust contamination was started immediately after the Seveso accident to assess inhalation risks of air-borne TCDD and to define the possible impact of air-transported TCDD-containing particles on nearby areas (Di Domenico et al., 1980). This monitoring included the sampling of both air-suspended and settling dust.

The sampling of air-suspended particles, whose dimensions ranged from 0.3 to 100 um, was carried out with high volume samplers equipped with glass fiber filters. Simulated sampling tests were performed in order to ascertain possible TCDD losses from suspended particles during sampling, using a dust of known TCDD concentration obtained from the Seveso Zone A soil surface. This test demonstrated the absence of significant TCDD losses.

The sampling of settling dust (dimensions ranging from about 10 um to approximately 500 um) was carried out with dust-fall jars.

In 1977, at the limits of the most contaminated part of Zone A, this monitoring indicated a suspended dust level in air slightly higher than 0.1 mg/m^3. The TCDD concentration in this dust ranged from < 0.17 to 0.5 ng/g. The TCDD amount per cubic meter of air ranged from < 0.02 to 0.06 pg/m^3 (the sampled suspended dust included the particles that can be inhaled).

The settling dust monitoring was continued for a few years, and the number of sampling points was increased from 6 to 16, in order to appropriately control the dust contamination in the three zones and assess the TCDD deposition rate per surface unit. In 1977 six sampling sites were controlled, five of which were allocated in Zone A or at its boundaries, and the last one in Zone B. During that year, detectable TCDD levels in settling dust were found only at Zone A's most contaminated site (average TCDD value: 0.26 ng/m^2; range of single monthly values: < 0.05 ng/m^2 - 0.75 ng/m^2) and, in Zone B, in a single monthly sample, at a site close to a municipal incinerator (TCDD level: 0.09 ng/m^2). In 1978, the range of the yearly averages of TCDD settling rates with dust at the 16 sampling sites was from <0.01 ng/m^2/day (corresponding to the analytical threshold) to 0.23 ng/m^2/day (value detected in the most contaminated area of Zone A). TCDD levels higher than the analytical threshold were found only at five sites (in Zone A and in the above mentioned site in Zone B). At the most contaminated site in Zone A, the single TCDD monthly average deposition rates ranged from 0.11 to 0.8 ng/m^2/day in the same period . At that site, the TCDD concentration in settling dust ranged from about one hundred ppt to about 2 ppb in the same time period.

The ratio between TCDD levels in settling dust and TCDD levels in the soil within some hundreds of meters from the dust sampling sites (0 - 7 cm deep soil layer) was generally between 0.1 and 1 (when measurable). A simple and conservative assessment indicated that the impact of the air-transported TCDD on soil TCDD levels could be considered practically negligible. In fact, even assuming an average dust settling rate on the ground of about 200-300 mg/m^2/day (i.e., in the order of the maximum monthly assessed settling rates) without any further movement of such dust from the involved soil (which is a very unlikely and conservative assumption), and a TCDD concentration in such dust of about 2 ppb (in the order of the maximum detected level in settling dust in Zone A), the dust deposition on soil during one year will be in the order of 100 g/m^2/year, with a yearly TCDD deposition in the order of about 0.2 ug/m^2/year. This level is well below the standard analytical threshold for soil standard monitoring and also below the TCDD soil levels detected in Zone R. If reference is made not to the maximum but to the average TCDD settling rate with dust, as measured in the whole Zone A, the same calculations indicate values at least 5-fold lower.

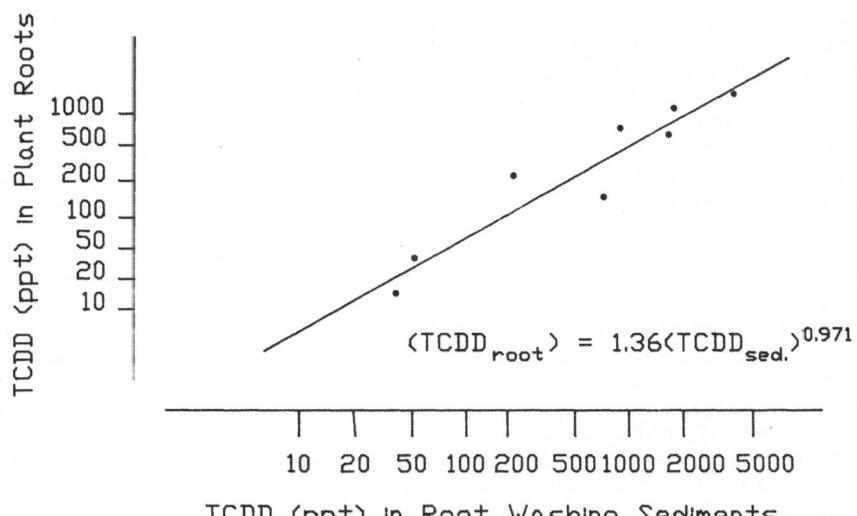

Figure 8.12 Regression of TCDD concentrations detected in maize and bean roots and in the sediments from root washing

The comparison of TCDD levels in dust samples from high-volume samplers and dust-fall jars indicated no substantial differences, therefore suggesting a minor or negligible influence of dust particle dimension on their contamination level.

Finally, the dust-fall jars data indicated a decreasing TCDD dust content for increasing distances from Zone A (Di Domenico et al.,1980a).

7   DISCUSSION AND CONCLUSIONS

The above discussed data show that the impact of environmentally released TCDD is mainly exerted on the soil compartment, as well as on plants and animals in direct contact with the contaminated soil. The data on TCDD vertical distribution in the soil of the Seveso area indicate that an initial significant TCDD permeation in the upper soil layers took place shortly after the soil contamination. After this "transitory process" the vertical mobility of TCDD appeared to decrease progressively. At about one year after the accident the TCDD vertical distribution in soil appeared nearly stable. The modelling procedures of chemical contaminant behaviour in soil, reported in this volume (C. Lupi) lead to predictions in agreement with these findings. The characteristics of the environmental release of TCDD during the accident, together with the rain action and TCDD's physical-chemical properties, can offer the key to explain this behaviour.

It can be reasonably assumed that immediately after the accident, which caused the environmental release of the TCDD-containing air-borne drops or particles, the TCDD contamination involved only the soil surface. Moreover it can be assumed that at that moment the TCDD was scarcely or not bound to soil organic carbon. Under such conditions, no significant obstacles existed for TCDD water transport; the only limit to TCDD's mobility with water was set by the TCDD solubility in water. Simple calculations indicate that at the most contaminated sites of the Seveso area the water present on the soil surface, as a consequence of the first rainfalls, was saturated by TCDD. This did not happen in medium or low level contaminated sites. Therefore, it can be reasonably assumed that the percentage of TCDD transported from soil surface to soil deeper layers was lower at highly contaminated sites (due to the saturation process) than at less contaminated sites (where the TCDD transport with water was not limited by solubility). This hypothesis can easily explain the differences observed in TCDD vertical distributions in soil at sites with different contamination levels.

After this "transitory process", it can be assumed that TCDD partitioning among soil-solids, soil-contained water and soil-contained air was considerably facilitated. "Environmental Partitioning" models (McCall, 1983; Mackay et al., 1982) indicate that, at equilibrium conditions, practically all the TCDD present in the soil is adsorbed to the organic carbon contained in soil solids, while its concentration in soil-contained water is extremely low (four orders of magnitude or more lower than that in soil solids). This is the obvious consequence of the very high soil adsorption constant of TCDD. Under such "steady state" conditions, the transport with water is extremely low or practically absent. This can easily explain the progressive decrease of TCDD's vertical mobility in soil after some months from its environmental release.

The environmental persistence of TCDD in the Seveso area was seen to be much higher than the one expected from laboratory experimental data on TCDD ultra-violet photodegradation. The limited efficiency of the ultra-violet solar degradation of soil-contained TCDD may be easily explained by the significant initial TCDD permeation in soil: at approximately five months after the accident, the statistical analysis of available data indicates that probably more than 60% of the TCDD present in soil was no longer available to UV degradation. The TCDD determinations in the very top soil layers confirmed the hypothesis of a UV degradation limited to the soil surface.

The indication of a higher efficiency of the degradation process in highly contaminated areas, arising from some of the data shown above, may be at least in part explained by the characteristics of the already discussed initial TCDD water transport process. In fact, at highly contaminated sites, the percentage of TCDD initially available to UV degradation may be assumed to have been higher than the one at medium and low-level contaminated sites, due to the limits set by TCDD water solubility to its percolation in soil with water.

The TCDD mobility with air-transported contaminated dust appeared to be relatively limited in the Seveso area; in any case, no significant changes of the TCDD distribution pattern in soil were attributable to this process.

The TCDD transfer from soil to plants in the Seveso environment was seen to be mainly due to water transport and to absorption of TCDD contaminated water by plant roots, as well as to soil-originating dust deposition on plant surface (and subsequent possible absorption). Laboratory data indicate that in closed environments a further possible plant contamination source can be represented by TCDD evaporation from soil. The soil-to-plant TCDD transfer appeared to be more efficient at medium and low level contaminated sites than at highly contaminated sites. Also in this case, the limit to TCDD water transport set by the TCDD solubility in water may be reasonably assumed to be one of the causes of the detected differences. In particular, several data from the Seveso area indicated a ratio between the TCDD concentrations in plants and in the corresponding soil higher at less contaminated sites than at highly contaminated sites (among other things, this indicated that simple extrapolations of this parameter from one class of sites to another were not fully reliable).

These considerations, together with some data on soil-to-animal transfer, indicated the necessity of a conservative approach in TCDD exposure evaluation.

The examination of all the data of the Seveso area, and of those obtained from laboratory studies carried out in connection with the Seveso accident, indicated a typical pattern of TCDD environmental contamination, characterized by a substantial stability and, therefore, implying a chronic or long term human and animal exposure. Inhalation, ingestion and dermal absorption of contaminated dust, ingestion of contaminated plant and animal foods are important exposure pathways.

In conclusion, these data showed that a comprehensive and joint consideration of all aspects involved, including mobility, environmental partitioning and distribution, degradation, transformation, accumulation and bio-accumulation processes, is essential for an adequate and complete description and modelling of TCDD's environmental behaviour.

8   REFERENCES

Adams, W.J., and Blaine, K.M., 1986, A water solubility determination of 2,3,7,8-TCDD, Chemosphere, 15:1397.

Cocucci, S., Di Gerolamo, F., Verderio, A., Cavallaro, A., Colli, G., Gorni, A., Invernizzi, G., and Luciani, L., 1979, Absorption and translocation of tetrachlorodibenzo-p-dioxins by plants from polluted soil, Experientia, 35:482.

Cocucci, S., 1980, "Absorption, translocation and elimination of TCDD by plants in contaminated soil", International Steering Committee 3rd Meeting, March 30-April 1, Segrate (Milano), Italy.

Di Domenico, A., Silano, V., Viviano, G., and Zapponi, G.A., 1980, Accidental Release of 2,3,7,8-Tetrachlorodibenzo-p-dioxin (TCDD) at Seveso, Italy, Ecotoxicology and Environmental Safety, 4:282.

Di Domenico, A., Viviano, G., and Zapponi, G., 1982a, Environmental persistence of 2,3,7,8-TCDD at Seveso, in: "Chlorinated Dioxins and Related Compounds. Impact on the Environment", O. Hutzinger, R. Frei, E. Merian, and F. Pocchiari, eds., Pergamon Press, Oxford.

Di Domenico, A. Viviano, G., and Zapponi, G., 1982b, Methodological problems in assessing 2,3,7,8-TCDD environmental contamination at Seveso, in: "Chlorinated Dioxins and Related Compounds. Impact on the Environment", O. Hutzinger, R. Frei, E. Merian and F. Pocchiari, eds., Pergamon Press, Oxford.

Esposito, M.P., Tiernan, T.O., and Dryden, F.E., 1980, Dioxin, EPA-600/2-80-197, Cincinnati, USA.

Facchetti, S., Blasso, A., Fichtner, C., Frare, G., Leoni, A., and Mauri, G., and Vasconi, M., 1984, "Studi relativi all'assunzione di TCDD da parte di alcune specie vegetali", Atti del Convegno "La risposta tecnologica agli inquinamenti chimici", September 20-22, 1984, Regione Lombardia, Milano, Italy.

Facchetti, S., Balasso, A., Fichtner, C., Frare, G., Leoni, A., Mauri, C., and Vasconi, M., 1986, Studies on the absorption of TCDD by some plant species, Chemosphere, 15:1387.

Hay, A., 1982, "The Chemical Scyte", Plenum Press, New York (USA).

Inquiry of the experts designated by the magistrate entrusted with judiciary proceedings on the Seveso accident, Tribunale di Milano, 1979.

McCall, P.J., Laskowski, I.A., Swann, R.L., and Dishburger, H.J., 1983, Estimation of environmental partitioning of organic chemicals in model ecosystems, Residue Reviews, 13:231.

Mackay, D., and Paterson, S., 1982, Fugacity Revisited, Env. Sci. and Techn., 16:654.

Pocchiari, F., Di Domenico, A., Silano, V., and Zapponi, G.A., 1983, Environmental impact of the accidental release of tetrachlorodibenzo-p-dioxin (TCDD) at Seveso (Italy), in: "Accidental Exposure to Dioxin. Human Health Aspects", F. Coulston and F. Pocchiari, eds., Academic Press, New York.

Pocchiari, F., Cattabeni, F., Della Porta, G., Fortunati, U., Silano, V., and Zapponi, G.A., 1986, Assessment of exposure to 2,3,7,8-tetracholorodibenzo-p-dioxin (TCDD) in the Seveso area, Chemosphere, 15:1851.

MATHEMATICAL AND BIOLOGICAL UNCERTAINTIES

IN THE ASSESSMENT OF A PERMISSIBLE BLOOD LEAD CONCENTRATION

F. Sartor and D. Rondia

Environmental Toxicology Unit, B6
University of Liège, Belgium

1   INTRODUCTION

One of the aims of environmental toxicology is to assess the human health risk created by the presence of pollutants in most compartments of our environment, e.g. air, drinking water, surface water, food, wastes etc.. Identifying and evaluating the health risk for humans is one of the most decisive steps in the development of environmental risk management.

Unlike clinical or drug toxicologists, environmental toxicologists do not use indexes of lethality (LD50 or LDLo, minimum lethal dose). On the contrary, their approach is based on the lowest detectable effect, be it adverse (NOAEL, no observed adverse effect level) or not (NOEL, no observed effect level). There is also a tendency to base these NOAELs or NOELs, as far as can be done, on human data rather than on animal data in order to avoid the uncertainty factors inherent in the extrapolation of animal experiments to human beings. This procedure has been recommended, when possible, by the World Health Organization (WHO, 1987). In the present study, we shall apply this approach to the assessment of a NOAEL for the exposure of a general population to environmental lead.

The pollution of the environment by lead has significantly increased since the industrial revolution. Two major geochemical evidences have demonstrated this fact:
- the significant increase of the concentration of lead in successive snow layers sampled from the Greenland inland ice (Murozumi et al., 1969) and
- the inversion of the natural gradient of the lead concentrations in seawater (Chow and Patterson, 1966).
The pollution of the atmosphere by lead plays an important role in the contamination of the global environment and the largest source of anthropogenic emissions of lead to the atmosphere originates presently in the combustion of gasoline lead. The global emission of lead from this source for year 75 was estimated at 60% of the total man-made release of lead to the atmosphere (Nriagu, 1979).

Environmental lead reaches man via inhalation and through the food chain after atmospheric deposition on plant foliages, soils or seawater, and uptake by crops through plant roots. The increase of the lead contamination in contemporary humans was demonstrated by Ericson et al.

(1979). They found a 500-fold elevation of the lead/calcium ratio in skeletons of present-day residents of England and the United States as compared to prehistoric values. Thence the influence of environmental lead exposure on the human health has been extensively investigated these last three decades and the results of many epidemiological surveys have supported the firm belief that the reduction or the suppression of gasoline lead would definitively ensure the protection of the human from undue lead uptake. Such strategy will only hold if the NOAEL does not change over time. As the NOAEL for lead exposure is regularly revised with a decreasing trend, this hypothesis is unrealistic and the effort to lower the atmospheric pollution by lead becomes inadequate.

During the last 15 years, our laboratory has been concerned with the toxicity of lead at low doses from air and drinking water pollution. Much of our work was done on the population of a small town near Liège where the drinking water is extremely soft and where the houses, until recently, were all provided with lead piping. This situation has resulted in a significant impregnation of the local population by lead (Sartor and Rondia, 1980). The source of impregnation can be considered as homogeneous despite the fact that persons or families have to various extent obeyed the official precautionary instructions. It provided us with a larger than usual span of blood lead concentrations, especially among children, and with a reference population having exactly the same living conditions (except the use of lead contaminated water). A second reference population was also sampled in the Liège area which is supplied with a very hard drinking water.

2   UNCERTAINTIES IN THE USE OF NOAEL

There is a general agreement among public health scientists to accept the lowest blood lead concentration (PbB) at which a significant increase of free protoporphyrins concentration in red blood cells is observed as the lowest adverse effect level of lead in humans.

The fact that the free erythrocyte protoporphyrin (FEP) concentrations do not decrease to zero as blood lead decreases, shows that there is a background FEP value for each of us owing to normal physiological reasons (e.g. presence of a few immature red cells in the circulating blood). The existence of this background complicates the mathematical expression of the relation between the two variables and is therefore the source of a mathematical uncertainty in the risk assessment of lead toxicity.

Furthermore, in some persons the physiological reasons for the presence of the FEP background are not exactly the same as in the majority of the population. This is especially true for children at school age, which is a period of marked growth in stature and in sexual maturity. On the other hand, children are often choosen as the critical population because they show the highest sensitivity to lead absorption. Such factors, completely unrelated to the lead concentrations in the environment, are consequently the source of a biological uncertainty of the NOAEL, to which insufficient attention has been paid until recently.

The NOAEL is equivalent to the permissible dose when its evaluation is based on the study of the dose-effect or the dose-response relationships in human populations. A dose-effect curve gives the relation between the dose and the magnitude of a graded effect, either in an individual or in a population. A dose-response curve shows the relation between the dose and the proportion of individuals responding with a quantal (binary) effect (WHO, 1978).

This report studies mathematical and the biological uncertainties in the sense of the model and parameter uncertainties described by Trønnes and Heiberg in Chapter 2. These uncertainties influence the results of the analysis of the dose-effect or dose-response relationships for a given critical effect (i.e. the particular effect producing an undesirable functional change, reversible or not, at the cellular level; Nordberg, 1976). The mathematical uncertainty will only be discussed from a theoretical point of view while the biological uncertainties underlying the relationship between blood lead and free erythrocyte porphyrin concentrations will be demonstrated by the analysis of epidemiological data collected in a group of children exposed to a wide range of environmental lead levels.

3   MATHEMATICAL UNCERTAINTY

In environmental risk assessment the mathematical uncertainty refers to different concepts: the precision or the imprecision of most statistical parameters (e.g. the 95% confidence interval of regression coefficients) and the probability of occurrence or non-occurrence of an event constitute two classical ways for conceiving uncertainty. In this report, we shall refer to mathematical uncertainty as the questionable confidence level of an estimated permissible dose resulting from the possible inadequate choice of statistical model.

For a long time, experimental toxicologists have extensively used probit analysis to fit curves between dose and frequency of a quantal response (binary effect or all-or-nothing effect). The median lethal dose, LD50, in animal experiments is usually estimated by this statistical procedure (Finney, 1971). In a probit analysis, the response is the prevalence of a specific binary effect expressed as a percentage i.e. the proportion of individuals being affected at a given dose. As the relation between the dose and the response has generally a sigmoid shape, it is necessary to linearize the curve in order to get the best estimate of its parameters by a linear regression line. This is achieved by a probit transformation of the observed responses which is defined as follows : if z (p) is the value of the standardized normal deviate corresponding to a response p, the probit y of p is given by

$$y = 5 + z(p)$$

The parameters of the regression line between y probit of the response and dose (or its logarithm) can not however be estimated by the classical least square method because the variance of y is not constant over the entire range of doses (Armitage, 1971). The iterative reweighted regression technique to be used in this case is fully described by Finney (1971).

It was natural that the same model would be employed to handle the epidemiological data collected to study the health effects of metallic pollutants. In environmental health studies, the response gives the frequency of the individuals showing a quantitative effect exceeding a reference level: e.g. the percentage of persons whose urinary δ-aminolevulinic acid concentration is above 5 mg/g creatinine at various blood lead levels (Figure 9.1). To justify this dichotomy, it is often argued that the permissible dose should be based on dose-response relationships rather than on dose-effect relationships because they take

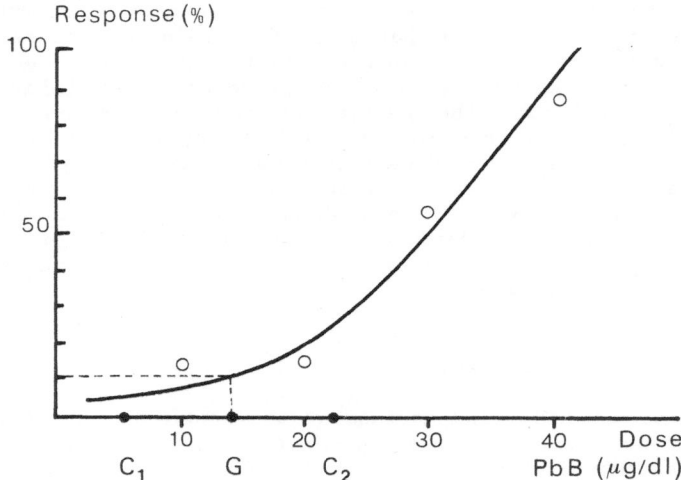

Figure 9.1 Assessment of a permissible dose using the probit analysis (o: the observed response; G is the permissible dose at which 10 % of the population would respond; C_1 and C_2 are the lower and upper 95 % confidence limits of G)

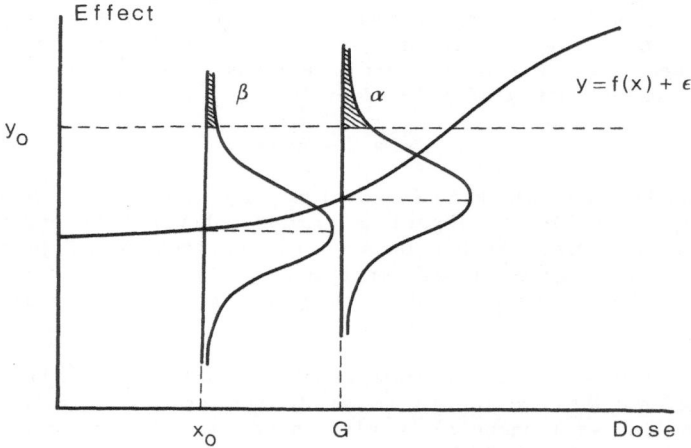

Figure 9.2 Assessment of a permissible dose using continuous dose-effect regression

better account of the differences in individual sensitivity resulting from the heterogeneity of the general population.

In our opinion (also shared by Whitehead, 1980), this method is unsatisfactory owing to the following drawbacks:
- the computation of the response for a given dose (frequency for a quantitative effect to exceed a reference level) results in a loss of information.
- the reduction of the quantitative effect into a binary or quantal effect requires the knowledge of a reference level for the investigated effect. This is frequently computed before the fit of the curve by considering the lowest exposed group as a control group. This is a good example of a vicious logical circle.
- there are no general rules for choosing the acceptable level of the response. In other words, has the permissible dose to be computed for a response of 1, 2, 5, 10 % or a greater value in the general population?
- from our experience in studying the dose-response relationships for blood lead, δ-aminolevulinic dehydratase and urinary δ-aminolevulinic acid in adults and for blood lead and free erythroporphyrins in children, it was concluded that too large 95% confidence intervals are found in estimating the doses at which 5 and 10 % of the population would respond, even for a large population sample (Sartor, 1988). This is a crucial point especially when the dose without adverse effects is very low.

The analysis of a continuous dose-effect relationship fitting non-reduced values of a quantitative effect constitutes a second approach to the mathematical assessment of a permissible dose. This evaluation requires the following data (Whitehead, 1980; Figure 9.2):
- the estimation of the regression coefficients of the function $y = f(x)$ by the classical least squares method. The adequacy of the model must also be verified by the analysis of the residuals,
- the determination of y_0, reference level of the graded effect observed in $100 \cdot \beta \%$ of the individuals exposed to the dose x_0 ($\beta = 0,025$ in general). The value of x_0 is chosen so that it is assumed to be a reference or control dose level,
- the choice of α, the acceptable probability of exceeding the reference level y_0 in the individuals exposed to the permissible dose G, i.e. the dose at which the magnitude of the graded effect is higher than y_0 in $100 \cdot \alpha \%$ of the individuals exposed to that dose ($\alpha = 0.10$ in general) and
- σ, the standard deviation of the error term ε in the model $y = f(x)$.

Whith these informations, one can easily compute the permissible dose G from the general equation:

$$f(G) = f(x_0) + \sigma \cdot (z_\beta - z_\alpha)$$

where z_β and z_α are the values of the standardized normal deviate for the probabilities β and α (these values are respectively 1.96 for $\beta = 0.025$ and 1.28 for $\alpha = 0.10$).

This analysis must (ideally) be conducted with data collected in homogeneous groups of the population in order to minimize the effect of the interindividual variability. For lead exposure, it is well known that children are at higher risk than adults. It is therefore reasonable to assess the permissible dose on the basis of the dose-effect relationship for the critical effect in this particular subgroup of the general

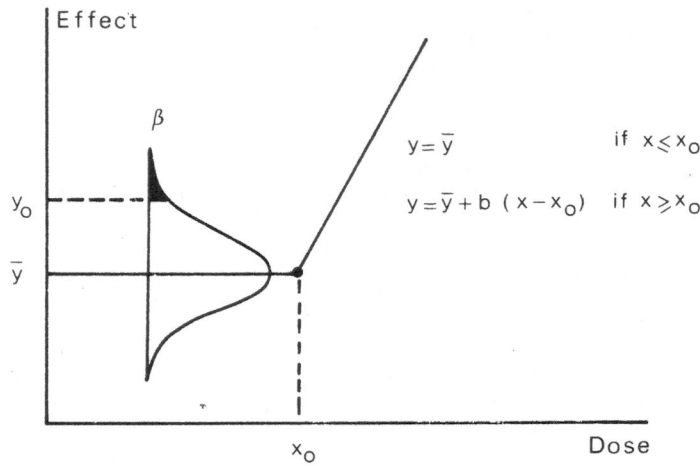

Figure 9.3   Assessment of a permissible dose
using a segmented dose-effect
regression

population. Such a method does, however, not prevent all the objections previously mentioned for the probit analysis (the influence of a chosen x_0 on the permissible dose G is obvious in the above mentioned equation).

Dose-effect and dose-response relationships may lastly be analyzed by the hockey stick regression (Figure 9.3). In contrast to the two preceding models, this segmented line function assumes that a concentration threshold below which no effects are discernable does exist. This underlying hypothesis may be questioned but has received considerable support, the demonstration of adaptive mechanisms being the most important. The main advantage of this method, developed by Hasselblad et al. (1976), is to estimate a threshold dose value x_0 without no a priori reference to a control group. It is furthermore possible to make hypothesis testing as the 95% confidence limits for x_0 can easily be computed. The goodness of fit for dose-response data has also been investigated by Hasselblad et al. (1976). There was little or no difference between the fit of the "hockey stick", the probit and the logit curves for various sets of data simulated from known probit (or logit) functions.

The algorithm of the computational procedure used in the hockey stick model is derived from the general theory proposed by Hudson (1966) to fit segmented curves whose join points have to be estimated. The overall residual sum of squares of a bi-linear model is partitioned in two separated residual sums of squares corresponding to each line of the model. If x_0 lies in an interval between two successive observed values of x, it can be demonstrated that it is sufficient to minimize the two residual sums of squares separately to minimize the overall residual sum of squares. For the hockey stick model, this theorem holds also if x_0 lies at one of the observed x values. The application of this theorem provides two sets of three normal linear equations in which x_0, b and $\bar{y}$ are the unknowns. It is thus easy to find the algebraic expressions giving

the values of x_0, b and $\bar{y}$ under the above hypothesis (these expressions are fully expanded in the report of Hasselblad et al., 1976). The computer program works as follows:
- the observed values of x are increasingly ordered,
- x_0, b, $\bar{y}$ and the overall residual sum of square are computed for each observed value of x and for each interval between two successive values of x,
- the values of x_0, b and $\bar{y}$ corresponding to the minimum overall residual sum of squares are selected and,
- they are used as starting parameters in the NLIN procedure (non-linear regression) of the SAS package (1982) to estimate their 95% confidence limits.

In conclusion, this review of the available statistical methods allowing the assessment of a permissible dose for metallic pollutants demonstrates that the hockey stick regression is probably the method minimizing the mathematical uncertainty in the decision making procedure. It will be used in the next section of our contribution.

4   BIOLOGICAL UNCERTAINTY

The assessment of the permissible dose for a given pollutant is also subjected to another kind of imprecision which is harder to express in mathematical or statistical terms. This uncertainty is essentially due to the specific biological features inherent in the investigated dose-effect or dose-response relationships. This so-called biological uncertainty does not only arise because of a lack of specificity in the investigated effect. The results of an epidemiological study we have carried out in a small urban center (Verviers: 50,000 inh. ; 30 km east of Liège) will demonstrate that other parameters, e.g. exposure duration, could have a great influence on the assessment of the permissible dose.

4.1 Material and Method

Many dwellings of the investigated area are supplied with a soft and corrosive drinking water through lead pipe systems. A significantly higher risk of lead uptake in the population of this area has been demonstrated by several epidemiological surveys (Sartor and Rondia, 1980; 1981). The prevalence of blood lead higher than 35 µg/dl is about 10% for the whole population of the Verviers area while it reaches 30-35% in the subpopulation living in houses with lead pipes. These frequencies exceed largely the recommended EEC guidelines: less than 2% of the blood lead concentration in population samples should be above 35 µg/dl (EEC Directive, 1977).

The eligible population, relevant to this report, included all the children (exposed and non-exposed to lead contaminated soft water) attending five elementary schools in the Verviers area (n=341). The formal consent of parents for collecting venous blood was obtained in 228 cases (participation rate: 67%). Blood lead (PbB) was determined by electrothermal atomic absorption spectrometry. Free erythrocyte porphyrin (FEP) was measured with the micromethod of Piomelli (1977) and expressed in µg/dl of red blood cells (RBC). Haemoglobin (Hb) was determined spectrophotometrically after conversion to cyanmethemoglobin and haematocrit (Ht: volume occupied by red and white blood cells, expressed as a percentage) with an IEC microcentrifuge. HbA2 component of haemoglobin was also determined by electrophoresis. Complete blood analysis was available for 204 children only.

Table 9.1 Results of the Blood Analysis, by Sex, in Children
          Aged 6-16 Years Living in the Verviers Area (82
          Boys and 122 Girls, Exposed and Non-exposed)

	Boys	Girls
age (years)	9.4 ± 2.2	9.6 ± 2.2
	(6 - 14)	(6 - 16)
PbB (µg/dl)	21.5 ± 6.9	20.8 ± 7.9
	(9 - 45)	(9 - 51)
Hb (g/dl)	13.1 ± 0.97	13.2 ± 1.00
	(10.2 - 14.9)	(9.5 - 15.5)
Ht (%)	41.2 ± 3.0	41.1 ± 3.14
	(29.5 - 47.7)	(33.9 - 49.4)
FEP (µg/dl RBC)*	38 ± 1.89	39.6 ± 1.98
	(12.5 - 200)	(6 - 658)

Values are arithmetic means ± standard deviations (the range
is in parenthesis); *: geometric mean ± standard geometric
deviation.

4.2 Results and Comments

    The results are summarized in Table 9.1. They show that:
- PbB are in the range 9-45 µg/dl for boys and 9-51 µg/dl for girls.
- the geometric means of the FEP are in the range of values reported by
  Piomelli (1977) for PbB levels around 20 µg/dl .
- the mean Hb concentration and the mean Ht agree with previously obtained
  values (Hb: 12.6-13.4 g/dl; Ht: 37.9-40%; Ciba-Geigy, 1972) but the
  extreme lower figures reveal a few cases of frank or mild anemia (Table
  9.1). Indeed, 11 children have a Hb concentration lower or equal to 11.5
  g/dl. It is therefore inferred that the prevalence of anemia is about 5%
  in the examined group if one accepts our Hb cutoff level as a criteria
  for anemia in children.

Table 9.2 First and Second Order Partial Correlation
           Coefficient between PbB, Hb, Ht and log FEP in the
           Examined Children Living in the Verviers Area

Correlated variables	Constant parameters	partial correlation coefficient	
		boys	girls
PbB-logFEP	age	0.324 ***	0.499 ***
PbB-Hb	age	0.256 *	0.005
PbB-Hb	age, logFEP	0.350 ***	0.140 *
logFEP-Hb	age	- 0.276 *	- 0.228 **
logFEP-Hb	age, PbB	- 0.392 ***	- 0.265

Level of significance: *:p < 0.05; **:p < 0.01; ***:p < 0.001

In order to elucidate the causes of these anemias - Pb exposure and/or other causes - we have computed the first and second order partial correlations of the biochemical parameters related to anemia (program 6R, BMDP Statistical Software, 1981). The results are shown in Table 9.2.

This analysis emphasizes the following facts:
- after removing the linear trend for age, there are significant <u>positive</u> correlations between PbB and Hb (in boys and at the 0.05 level only), and between PbB and log FEP. For both sexes, there is also a significant <u>negative</u> correlation between Hb and log FEP.
- the partial correlation between PbB and Hb in boys increases when both age and log FEP are kept constant. It still remains non-significant in girls.
- the <u>negative</u> correlation between Hb and log FEP becomes highly significant after removing the effect of age and PbB.

A negative partial correlation between PbB levels and Hb concentrations cannot be demonstrated in our examined group and the hypothesis of Pb induced anemias can therefore be excluded. Low haemoglobin concentration (hypochromia) and lower red blood cell volume (microcytosis) may also occur with a minor ß-thalassemia. This inherited disease is characterized by a lack in the production of haemoglobin and an imbalanced synthesis of the various proteic components of the haemoglobin. It is recognized by an abnormal increase of the A2 type haemoglobin expressed as a percentage of total haemoglobin. The frequency of this genetic disorder is especially high among Mediterranean people (Calabrese, 1978). Mediterranean by birth children make up 52% of the group examined in Verviers but the potential confounding effect of a minor ß-thalassemia may be excluded as no HbA2 component of haemoglobin exceeds the normal range (1.5-4%).

Nutritional iron deficiency, especially in women and children, is not uncommon even in developed countries (Cook, 1982). It therefore appears as the probable source of the observed anemias. The frequency of iron deficient erythropoiesis in apparently healthy children may not thus be neglected in assessing a PbB threshold based on the FEP response. A first source of non measurable uncertainty arises when we do not pay attention to this interference: we are not able to predict how and to what extent the dose-effect curve would be modified if all cases of mild iron anemia were removed.

The concentration of FEP depends simultaneously on the lead and on the iron body burdens. It is also well known that an insufficient iron absorption from the gut increases the gastro-intestinal lead uptake. In addition, our figures indicate that the relationship between FEP and Hb does not necessarily depend on the PbB level. A second source of non measurable uncertainty lies therefore in the biological interdependencies of these parameters and particularly in the fact that these dependencies are not approached in the same way by all environmental scientists.

In our sample, the PbB threshold value was estimated at 24.3 μg/dl by the hockey stick regression (95% confidence limits: 21.9 and 26.6 μg/dl; Figure 9.4). This estimate does not depend on the point corresponding to the highest PbB and FEP observed in our sample (PbB = 51 μg/dl and FEP = 658 μg/dl RBC). After removing this observation from the computation, the value of the PbB threshold is 24 μg/dl (95% confidence limits: 21.2-26.8 μg/dl). When the PbB level is lower than the threshold value, the geometric mean, $\bar{y}$, and the value of the 97.5th percentile, y_o, of the FEP distribution predicted by the model are respectively 33 and 91.2 μg/dl RBC. These figures agree with the values computed in the sub-group of the

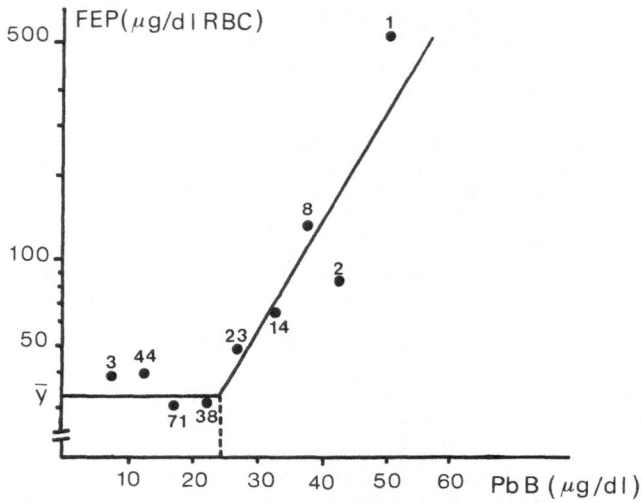

Figure 9.4　PbB　threshold　estimated
in　our　study.　Dots　are
geometric　means;　values
give　the　sample　size　for
each　PbB　interval　(0-10;
11-15;　16-20,　etc.)

156 children with PbB lower than 25 µg/dl: 33.1 and 80.5 µg/dl RBC.

      Our estimated PbB threshold level is higher than the figure reported
by Piomelli et al. (1982): 18.3 µg/dl (95% confidence limits: 17.1 and
20.2 µg/dl). At these PbB levels, a difference of 6 µg/dl (25% compared to
our estimate) could not be regarded as a negligible one. After a careful
examination of the two studies (Sartor, 1988), it was concluded that this
difference:
- is independent of the units used to express the FEP concentration (µg/dl
  whole blood, µg/dl RBC and µg/g Hb),
- could not be attributed to measurement errors in blood lead or FEP
  determinations,
- could be fully explained by the higher slope for the second segment of
  our fitted hockey stick function (see Figure 9.5) . This fact led us to
  investigate the influence on the FEP level increase of an unusual lead
  accumulation in the bone compartment. Such lead accumulation depends to
  a large extent on the lead exposure duration if the exposure level is
  assumed to be constant. In our group, the mean residence time in the
  dwelling at the moment of the survey was 4.5 years, i.e. about half the
  mean age in the examined group (Table 9.1). Children having PbB greater
  than 25 µg/dl have probably been exposed for several years to this
  unusual lead contamination. Some exclusion criteria used by Piomelli et
  al. (1982; suspicion of Pb poisoning, chelation therapy and PbB above 40
  µg/dl at any previous time) indicate that children with acute or chronic
  lead poisoning were discarded from their analysis. In addition the mean

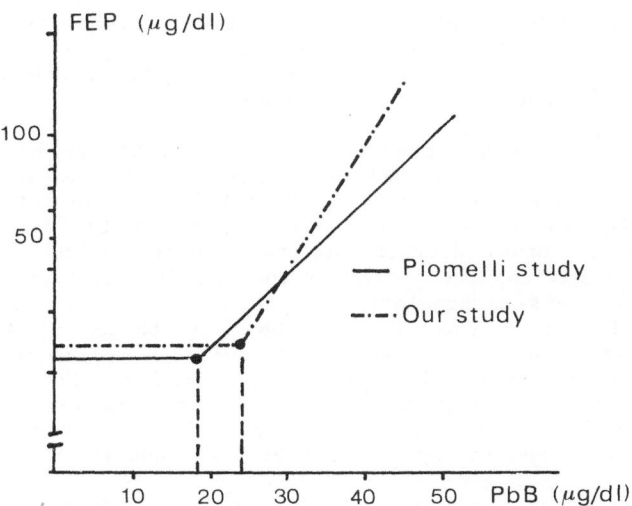

Figure 9.5 Influence of the slope of the hockey stick function on estimation of the PbB threshold value

age in their final study group (2,004 chidren aged 2-12 years) was 5.5 years, i.e. half the mean age of our group. It was therefore inferred that the lead deposits in the body of the children studied by Piomelli et al. were probably lower than in our sample owing to a shorter mean duration to current lead exposure. The hypothesis about the influence on the FEP concentrations of the lead deposits in the body is also supported by the findings of Alessio et al. (1976; 1981): at the same PbB levels, the FEP concentrations were more elevated in workers who had stopped working with inorganic lead for at least one year as compared to currently exposed workers. The FEP concentration may consequently stay at a high level due to a direct inhibition of haem synthesis by the lead released from the deposits. The relatively low PbB concentrations, as opposed to the high FEP in subjects no longer exposed, could result from the fact that the amount of lead diffusing in the blood from the deposits is small and is therefore not able to influence the PbB concentrations.

5   FROM RISK ASSESSMENT TO RISK MANAGEMENT

Despite the slight difference between Piomelli's figures and our results, and the bias possibly introduced in the FEP concentrations by the biological iron reserve at very low levels of blood lead, one can say that our study confirms the American work and that both figures are based on human experience. According to this most valuable information, the NOAEL for environmental lead exposure should not exceed 20 µg/dl, a threshold value rounded off from Piomelli's and our study. This estimation depends obviously on the choice of the critical effect. Early heme synthesis damage is generally considered as the critical effect for humans but some scientists claim that other adverse effects, especially minimal neuropsychological disturbances, could occur at PbB levels below 20 µg/dl. Their causality is however still in discussion.

There is a general agreement to assume that the highest PbB value reasonably observed in a children population corresponds to the 98th percentile of the PbB distribution. This value should not be higher than the estimated NOAEL to ensure safe PbB based biological guidelines. Gaussian and lognormal distribution based calculations show us that to fulfil this condition, the median of the PbB distribution should be about 8-10 µg/dl in the same population. To arrive at this result we assumed a standard deviation in the Gaussian PbB distribution of 4.6 µg/dl. This figure was estimated in a sample of 126 children living in the Liège area where the environmental lead exposure is normal for Belgium (range: 5-25 µg/dl; mean PbB = 13.3 µg/dl). For the lognormal law, we have used a geometric standard deviation of 1.1 ug/dl. This value is slightly lower than the figures reported by Vather (1982) for adults: 1.3-1.6 µg/dl.

With these PbB based biological guidelines, we see from Figure 9.5 that the risk for the public health will be negligible. The occurence rate of a moderate increase in the FEP concentration amounts to 0.0005. It is computed by summing up the products of the probability density of PbB at various levels (20, 21, 22, etc. µg/dl) and the probability to observe a FEP concentration higher than y_0 (91.2 µg/dl RBC, i.e. the 97.5th percentile of the reference FEP distribution) at each considered PbB level.

We want to introduce these figures in decision-making about atmospheric pollution by lead and to establish an air quality guide for lead. We may use a model established by Snee (1981) from epidemiological data published in the literature. He shows that, in man, a decrease in atmospheric lead of 1 µg/m³ air results in a decrease of 1 ± 0.3 µg Pb/dl in whole blood (0.9 ± 0.4 µg/dl for women and 1.3 ± 0.3 µg/dl for children). Median PbB in Belgian adults living in Brussels and Liège was measured in 1981 by Claeys-Thoreau et al. (1983). It amounts respectively to 15.6 and 13.9 µg/dl.

The daily average levels of the major heavy metals were routinely determined in air samples collected during the period 1972-1977 at fifteen different monitoring stations located in industrial and non-industrial Belgian towns as well as in rural areas (Kretzschmar et al., 1980). This nationwide network was intended to assess the average quality of the air to which the Belgian citizen was exposed. Over the investigated period, the yearly lead averages were in the range 0.2-1.2 µg/m³. The time-series analysis of these data has not yet been performed in order to evidence a possible temporal trend of the mean daily lead concentrations in air.

From the above one may conclude that lowering atmospheric lead by lowering gasoline lead to zero will not bring a great part of the population under the safe median limit of 10 µg Pb/dl whole blood. From 0.8 g Pb per liter gasoline in 1970, we are presently down to 0.15 g/l, while blood lead in blood donors living in the Liège area has lowered from a 20 µg/dl average in 1977 to 13.9 µg/dl in 1981 (Sartor, 1988; Claeys-Thoreau et al., 1983). This significant decrease may not be fully explained as an immediate effect of the reduction in gasoline lead if we refer to the figures previously mentioned for the Belgian yearly lead averages in air and for the air-lead/blood-lead relation. The magnitude of this trend suggests therefore that other causes of exposure have perhaps decreased independently from the air pollution by lead (e.g. the dietary habits in the general population or food quality may have changed over the last ten years).

The statement set up by EPA (1984), that blood lead levels are directly related to the amount of lead in gasoline will not hold for low blood lead figures. The background of the lead pollution will become of increasing importance with a decrease of this principal source of lead exposure. The sources of this background lead exposure can be found in food, canned food, earthware, plastics, etc.. Among children, toys may play a significant role and among the very young, the quality of food and prepared foods is probably most important. The non-softened hard drinking waters are unimportant sources of lead exposure as the concentrations of lead in these waters are usually very low (10 µg Pb/1 or less). For an infant 6 month old, living in Liège, the atmospheric lead intake is negligible (less than 0.5 µg/kg/day assuming a yearly average of 1 µg Pb/m^3 air) whereas the dietary lead intake is estimated at 10.5 µg/kg/day (Sartor, 1988). This figure agrees with the metabolic studies reported by Ziegler et al. (1978) who have demonstrated that the lead intake ranged from 0.83 to 22.6 µg/kg/day with a mean of 9.43 µg/kg/day in normal infants, younger than 2 years, fed with milk or formula and commercially strained food. When the dietary lead intake is 8 to 9 µg/kg/day, blood lead concentrations increase with age whereas with a mean dietary intake of 3 to 4 µg/kg/day they do not (Ryu et al., 1983). The daily permissible intake for lead in infants should therefore be close to 3 µg/kg/day. For a 7 kg infant, the intake of lead, from diet alone, should thus be lower than 21 µg/day. At the present time, this figure is probably exceeded among most urban and rural young children.

Part of the dietary lead comes obviously from the atmosphere after deposition on soils and plants. A french study (Servant and Delapart, 1981) based on lead isotopic ratio estimates that atmospheric pollution accounted in 1977 for 20% only of the lead body-burden (6% from the pulmonary absorption and 14% from intestinal absorption via the food chain). With the same method, Petit (1983) showed that environmental pollution by lead increased by 50 percent since 1900 to 1950 in a remote place in Belgium and remained stable in quantity until the last years but with a slight increase in the 206 Pb/207 Pb ratio, indicating the growing proportion of gasoline lead in the environmental lead. At the present time, gasoline lead in exhaust gases does not probably play a major role in the overall lead contamination of human food because (Zimdahl and Arvik, 1973; Zimdhal, 1976):
- lead can be taken up by plant roots if present in a available form, but soil adsorption phenomena and the formation of insoluble compounds prevent uptake of large amounts of lead by plants,
- lead can be absorbed by plant foliage but much of it is apparently in the form of a topical deposit, and
- when taken up by a plant, lead may be found principally in the roots with little translocation to the shoot.
Furthermore, the highest figures found by Ziegler et al. (1978) for lead in infant foods (see above) univocally demonstrate the existence of other major sources of food contamination by lead. A recent paper on the progress made in quantitatively assessing the effects of low lead exposure on selected aspects of prenatal and postnatal development in children has resulted in similar conclusions (Davis and Svendsgaard, 1987).

Consequently, if the management of risk due to lead was to be based on health effects solely, these other known causes of lead impregnation of human being should, from now, receive as much attention and the same drastic solutions as those in application for gasoline lead.

6   ACKNOWLEDGEMENTS

We are indebted to NATO-CCMS for the CCMS Fellowship awarded to one of us for working in association with the Pilot Study on Risk Management of Chemicals in the Environment.

7   REFERENCES

Armitage, P., 1971, "Statistical methods in medical research", Blackwell Scientific Publications, Oxford.

Alessio, L., Bertazzi, P.A., Monelli, O., and Toffoletto, F., 1976, Free erythrocyte protoporphyrin as an indicator of the biological effect of lead in adult males, Int. Arch. Occup. Environ. Health, 38:77.

Alessio, L., Castoldi, M.R., Odone, P., and Franchini, I., 1981, Behaviour of indicators of exposure and effect after cessation of occupational exposure to lead, Brit. J. Ind. Med., 38:262.

BMDP Statistical Software, 1981, ed. by W.J. Dixon, University of California Press, Berkeley.

Calabrese, E.J., 1978, "Pollutants and high-risk groups. The biological basis of increased human susceptibility to environmental and occupational pollutants", J. Wiley & Sons, New York.

Chow, T.J., and Patterson, C.C., 1966, Concentration profiles of barium and lead in Atlantic Waters off Bermuda, Earth and Planetary Sciences Letters, 1:397.

Claeys-Thoreau, F., Bruaux, P., Ducoffre, G., and Lafontaine, A., 1983, Exposure to lead of the Belgian population, Int. Arch. Occup. Environ. Health, 53:109.

Ciba-Geigy, 1972, Scientific tables, Bâle, p. 628.

Cook, J.D., 1982, Clinical evaluation of iron deficiency, Seminars in Hematology, 19:6.

Davis J.M., and Svendsgaard, D.J., 1987, Lead and Child Development, Nature, 129:297.

EPA, 1984, "Risk assessment and management : framework for decision making", United States Protection Agency, report 600/9-85-002, p. 33.

EEC, 1977, Council Directive on biological screening of the population for lead (77/312/EEC), Official Journal of the European Communities, L105/10-17, 28 april 1977.

Ericson, J.E., Shirahata, H., and Patterson, C.C., 1979, Skeletal concentrations of lead in ancient Peruvians, N. Engl. J. Med., 300:946.

Finney, D.J., 1971, "Probit analysis", Cambridge University Press, Cambridge.

Hasselblad, V., Creason, J.P., and Nelson, W.C.,1976, "Regression using "Hockey stick" functions", EPA report 600/1-76-024, Research Triangle Park.

Hudson, D.J., 1966, Fitting segmented curves whose join points have to be estimated, Journal of the American Statistical Association, 61:1097.

Kretzschmar, J.G., Delespaul, I., and De Rijck, Th., 1980, Heavy metals levels in Belgium: a five years study, Sci. Tot. Environ., 14:85.

Murozumi, M., Chow, T.J., and Patterson, C., 1969, Chemicals concentrations of pollutant lead aerosols, terrestrial dusts and sea salts in Greenland and Antartic snow strata, Geochimica et Cosmochimica Acta, 33:1247.

Nordberg, G.F., 1976, "Effects and dose-response relationships of toxic metals", Elsevier, Amsterdam.

Nriagu, J., 1979, Global inventory of natural and anthropogenic emissions of trace metals to the atmosphere, Nature, 279:409.

Petit, D., 1983, Lead isotopes as stable tracers of atmospheric pollution in the environment, in: "Heavy Metals in the Environment", Intern. Confer. Heidelberg, CEP Consult., p. 148.

Piomelli, S., 1977, Free erythrocyte porphyrins in the detection of undue absorption of Pb and of Fe deficiency, Clin. Chem., 23:264.

Piomelli, S., Seaman, C., Zullow, D., Curran, A., and Davidow, B., 1982, Threshold for lead damage to heme synthesis in urban children, Proc. Natl. Acad. Sci. USA, 79:3335.

Ryu, J.E., Ziegler, E.E., Nelson, S.E., and Fomon, S.J., 1983, Dietary intake of lead and blood lead concentration in early infancy, Am. J. Dis. Child, 137:886.

Sartor, F., and Rondia, D., 1980, Blood lead levels and age: a study in two male urban populations not occupationally exposed, Arch. Environ. Health, 35:110.

Sartor, F., and Rondia, D., 1981, Setting legislative norms for environmental lead exposure: results of an epidemiological survey in the east of Belgium, Toxicology Letters, 7:251.

Sartor, F., 1988, "Evaluation des limites acceptables de l'imprégnation saturnine humaine. Contribution d'une enquête épidémiologique", Ph. D. thesis, University of Liège.

SAS (Statistical Analysis System) user's guide: Statistics, 1982, Cary, NC.

Servant, J., and Delapart, M., 1981, Blood Lead and Lead-210 origins in residents of Toulouse, Health Physics, 41:483.

Snee, R.D., 1981, Evaluation of studies of the relationships between blood lead and air lead, Int. Arch. Occup. Environ. Health, 48:219.

Vather, M., 1982, "Assessment of human exposure to lead and cadmium through biological monitoring", National Swedish Institute of Environmental Health and Department of Environmental Hygiene, Karolinska Institute, Stockholm.

Whitehead, J., 1980, The establishment and interpretation of dose-effect relationships for heavy metals pollutants, in: MARC report number 18, Chelsea College, London.

WHO Regional Office for Europe, 1987, "Air quality guidelines. Part I. General", WHO regional publications, European series, n° 23, Copenhagen.

WHO World Office, 1978, "Principles and methods for evaluating the toxicity of chemicals. Part 1", Geneva.

Ziegler, E.E, Edwards, B.B., Jensen, R.L., Mahaffey, K.R., and Fomon, S.J., 1978, Absorption and retention of lead by infants, Pediat. Res., 12:29.

Zimdahl, R.L., and Arvik, J.H., 1973, Lead in soils and plants: a litterature review, CRC Critical Reviews in Environmental Control, 3:213.

Zimdahl, R.L., 1976, Entry and movement in vegetation of lead derived from air and soil sources, J. Air. Poll. Control Ass., 26:655.

USE OF FORMAL METHODS IN EVALUATING COUNTERMEASURES TO COASTAL WATER
POLLUTION. A CASE STUDY OF THE KRISTIANSAND FJORD, SOUTHERN NORWAY

A. Heiberg and K.-G. Hem

Center for Industrial Research
Oslo, Norway

1   INTRODUCTION

The environmental problems occurring in modern industrial societies
are frequently characterized by a high degree of complexity. Often the
effects of pollutants released into the environment are uncertain and
difficult to assess. Moreover, the pollution may affect widely different
areas such as the natural environment, human health, outdoor recreation,
and industrial activities. The diversity of the pollution-induced effects
normally complicates the decision-maker's task of setting priorities and
choosing among available control options to a considerable extent. Among
the issues to be addressed is how to balance the various impact areas and
how to compare the benefits of pollution control with the cost of putting
an action into effect. Undoubtedly, methods or procedures that could
facilitate the handling of such problems would be of great help in the
decision-making process.

The basic elements involved in an analysis of pollution abatement
measures were discussed in Chapter 1. As will be recalled (see Figure
1.2), the analysis falls into two main parts, one in which the actual
measurable consequences of implementing the available control options are
examined, and one in which the significance of the different consequences
and their relative importance is assessed. In Chapter 1 an overview was
given of some formal methods that could be used in dealing with the second
task, the valuation of effects. The present chapter is concerned with the
use of formal methods in the context of coastal water pollution. Two
different valuation methods--economic analysis based on willingness-to-pay
surveys and the Simplified Multiattribute Rating Technique (SMART)--have
been explored through application to one particular case, the pollution of
the Kristiansand fjord, southern Norway. The prime purpose of the study
has not been to find realistic, recommendable measures for abating the
pollution of the Kristiansand fjord, but rather to gain experience with
employing formal decision-analytical methods in the management of
environmental risk. It is hoped that this experience will help establish
practical tools that will improve decision-making in the environmental
sector. For a more detailed account of the study we refer to Heiberg et
al. (1987) and to Tollefsen et al. (1988).

A number of areas along the Norwegian coast are heavily impacted by pollution. Among these is the Kristiansand fjord which is situated at the southeast part of the coast (see Figure 10.1). Unlike most of the other contaminated coastal areas in Norway, the Kristiansand fjord is located in a relatively densely populated area. At the head of the fjord is the city of Kristiansand with about 60,000 inhabitants.

2.1 Pollutant Loadings

The Kristiansand fjord receives wastewater from a variety of sources. The pollution consists primarily of organic matter, heavy metals (iron, nickel, copper, and cobalt in particular), organochlorine compounds (mostly chlorinated alkylbenzenes of different kinds), and polycyclic aromatic hydrocarbons (PAH). A major part of the pollution derives from the following sources:

2.1.1 Municipal Sewage. The input of municipal sewage from the Kristiansand region (mostly from households) totals about 82,000 person equivalents (pe's). Of this amount approximately 40,000 pe's are discharged into Vesterhavn and Fiskåbukta, the most strongly polluted part of the Kristiansand fjord (see Figure 10.1). Municipal effluents are the main source of the particulate organic matter found in this area. Most of the sewage is discharged into the receiving water without any special treatment or subsequent only to a simple mechanical cleaning.

2.1.2 Factories with Effluents Flowing into the Fjord. Wastewater from a nickel manufacturing plant located on the west side of the fjord, Falconbridge Nikkelverk (see Figure 10.1), has contributed a major portion of the input of metals and organochlorines. At present the daily discharge of iron, nickel, and copper from the factory is about 100, 50, and 20 kg, respectively.

2.1.3 The River Otra. Otra, which enters the sea at the city of Kristiansand (see Figure 10.1), receives loadings from agriculture, households, and industry. The pollutants include dissolved and particulate organic matter (fibres, fungi) attributable mainly to two wood-processing plants and to municipal sewage. One of the wood-processing plants also releases unknown amounts of acids and organochlorines into the river.

2.1.4 Benthic Sediments in Vesterhavn-Fiskåbukta. To a large extent the heavy metals, metal sludge, and organic micropollutants that have been discharged into Vesterhavn-Fiskåbukta over the past decades have accumulated in the bottom sediments. These sediments therefore constitute a reservoir of contaminants with the potential of releasing pollutants into the water body. Thus, even though the input from the industry has been significantly reduced over the last 5 years, the loadings of metals and organics will be substantial for many years to come unless special measures are taken. Today the benthic sediments are the main source of PAH in the area.

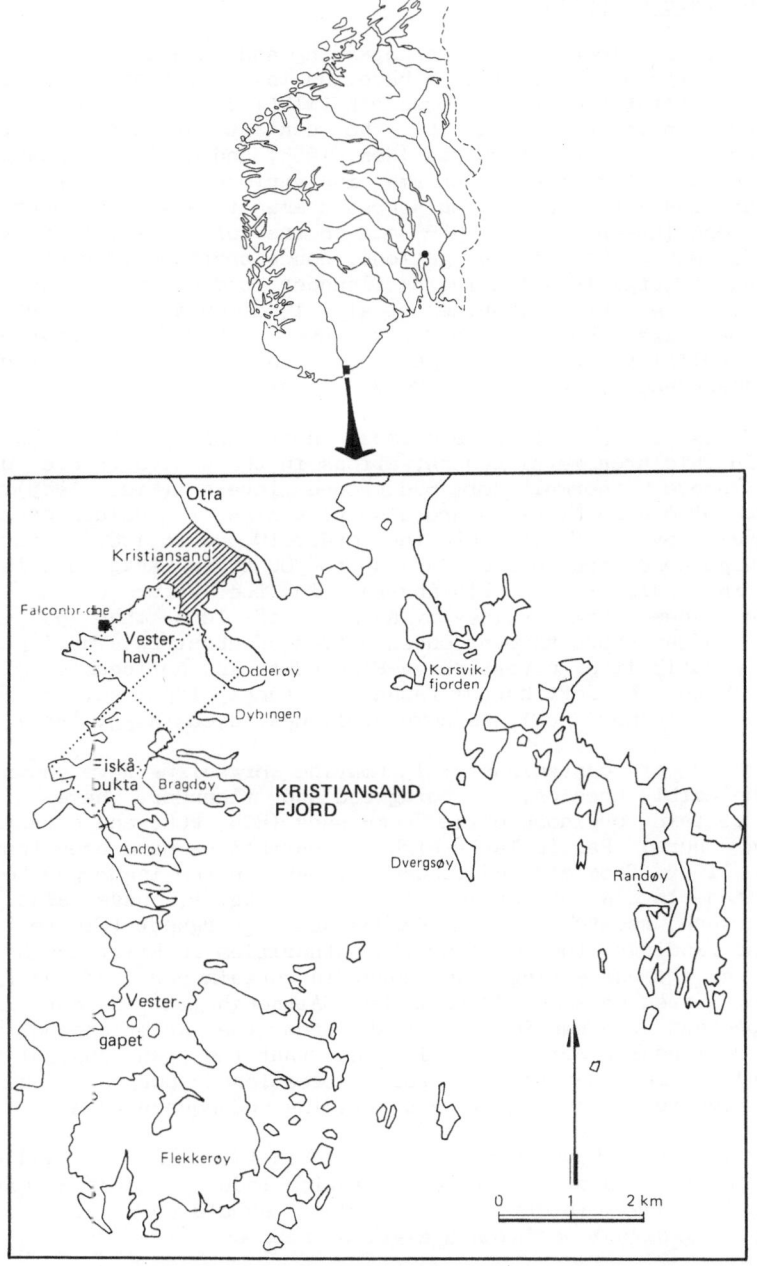

Figure 10.1   The Kristiansand fjord

2.2 Observed State of Pollution

Most of our knowledge about the spreading and impact of pollution in the Kristiansand fjord derives from a baseline investigation conducted by the Norwegian Institute for Water Reasearch (NIVA) during the period 1982-85. Measurements have been made of the content of contaminants in sediments and in the water column (Næs, 1985) and of the abundance and number of species of marine organisms in various seabed areas. The biological studies have focused on seabed communities in the rocky tidal and subtidal zone (Green et al., 1985) and on the soft bottom fauna (Rygg, 1985). In addition to these investigations, concentrations of contaminants in a wide range of organisms (including flounder, cod, common mussel and crab) have been measured (Knutzen et al., 1986, Knutzen and Martinsen, 1986). The baseline investigation also has provided information on temperature, salinity, content of phosphorus, nitrogen, and chlorophyll, and water transparency in the fjord (Molvær et al., 1986).

The effects of pollution are most noticeable in Vesterhavn and Fiskåbukta. In this area metal concentrations in the sediments are up to 800 times above "normal" or background levels (Næs, 1985). The concentrations of organochlorines are also very high, the values exceeding "normal" levels by a factor of the order 10,000 to 100,000 for some components. High concentrations of PAH, up to 800 times background levels, are found in sediments in Fiskåbukta. In the water column metal concentrations exceed normal values by a factor of less than 5, except with nickel which shows an overconcentration of about 20. The impact of pollution on aquatic life in Vesterhavn-Fiskåbukta is noticeable. In an area of 3 to 5 km^2 the soft bottom fauna is strongly impaired, and in the area closest to Falconbridge it is completely destroyed (Rygg, 1985).

Measurements of contaminants in marine organisms have shown that several of the edible species, including cod and flounder, have a high content of certain organochlorine compounds (HCB, PCB, and octachlorostyrene among others). Particularly high concentrations have been found in fillets and liver from cod and flounder caught in the inner part of the fjord. Elevated levels of organochlorines, however, have also been measured at such remote places as Randøy and Vestergapet (Figure 10.1). Owing to these findings, the local health authorities in Kristiansand have issued warnings against eating fish caught in certain parts of the fjord, covering a total area of about 20 to 25 km^2. Although it is not likely that the present consumption of fish from the Kristiansand fjord represents a real health problem, it is clear that the fish contamination makes the fjord less attractive for recreational purposes and for industrial activities such as commercial fishing and aquaculture.

In view of the above, one may conclude that the Kristiansand fjord represents a fairly complex picture of pollution-induced adverse effects and conflicting user interests. The fjord should be well suited, therefore, for trying out decision-analytical methods.

3 ANALYSIS OF COUNTERMEASURES - INITIAL TASKS

3.1 Effects to be Considered - Structuring of the Abatement Problem

As indicated in Figure 1.2 (Chapter 1), an analysis of pollution control actions involves identifying the types of effects to be considered and defining the variable to be used to describe each individual effect area. An important aim of this work, which must be carried out early in the analysis, is to obtain as complete a description of the problem as possible without overdescribing it. An efficient way to achieve this goal is to construct a hierarchy of objectives whose fulfillment is desirable. This also provides a structuring of the abatement problem which may be used in the evaluation of countermeasures (Keeney and Raiffa, 1976).

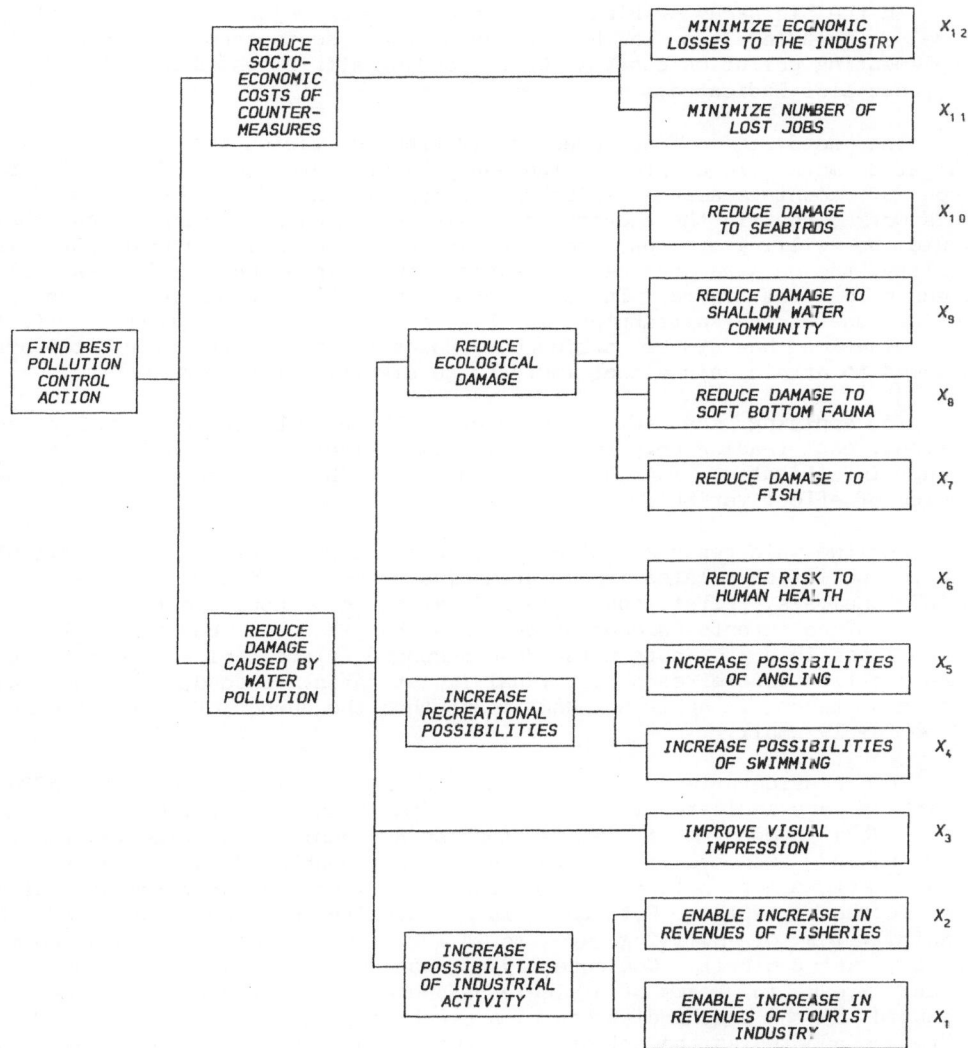

Figure 10.2 Objectives hierarchy for abating the pollution in Kristiansand fjord

An objectives hierarchy for the Kristiansand fjord case is shown in Figure 10.2. At the highest level of the hierarchy (far left) is the general, overall objective, formulated as "find the best pollution control action." This objective is split into a number of increasingly more specific subobjectives until, at the lowest level (far right), the objectives have become so precise that one can assign to them quantitative variables that indicate to what extent the objectives are fulfilled. If it can be agreed upon that these variables provide a fairly exhaustive description of the problem in question, they may be an appropriate set of effect parameters to use in the analysis.

3.2 Definition of Attributes

In the hierarchy of Figure 10.2 there are 12 objectives at the bottom level, including two which represent the socioeconomic costs of implementing pollution control. Corresponding attributes, designated X_1 to X_{12}, are also indicated.

The need to define precise attributes varies according to which valuation method is adopted in the analysis or, more correctly, to what type of countermeasures is being studied. In one of the methods used in this work , that is the economic analysis, one asks people how much they would be willing to pay for getting a "clean" Kristiansand fjord. In reality, the measure considered in this inquiry is a more or less full cleanup of the fjord. In this case there is no need to give precise definitions of the attributes in Figure 10.2 as no detailed effect determination has to be carried out. However defined, the attributes are assumed to attain values that indicate no significant adverse effect.

In applying the second method, SMART, a selection of more limited actions, each leading only to a partial improvement of the environment, is being investigated. In this case one has to be more specific about the choice of effect variables.

For two main reasons, it was considered impractical to include all of the ten pollution-related impact areas indicated in Figure 10.2 in the SMART analysis. First, our present state of knowledge regarding the effects of pollutants on ecosystems means that it would be difficult or even impossible to obtain the dose-response relationships needed for an adequate treatment of each individual impact area. Second, the special approach taken in applying SMART meant that the number of attributes had to be fairly small.

The considerations made to arrive at a smaller set of effect variables are indicated in Figure 10.3. Preliminary inquiries (Heiberg et al., 1987) showed that neither tourism nor commercial fisheries in the Kristiansand area are likely to suffer noticeable damage due to the present state of pollution. Therefore, the local industry was not given any further attention. Both the visual amenity of the fjord and the possibilities of swimming are to a large extent dependent on a common factor, water clarity. Consequently, it was decided to describe these two impact areas in terms of a single parameter, the water transparency, as measured by the secchi-disk depth. A secchi-disk depth of a minimum of 3 metre was adopted as the criterion of sufficiently high water quality. The attribute (x_1') was defined as the area of that part of the fjord where this criterion is not met.

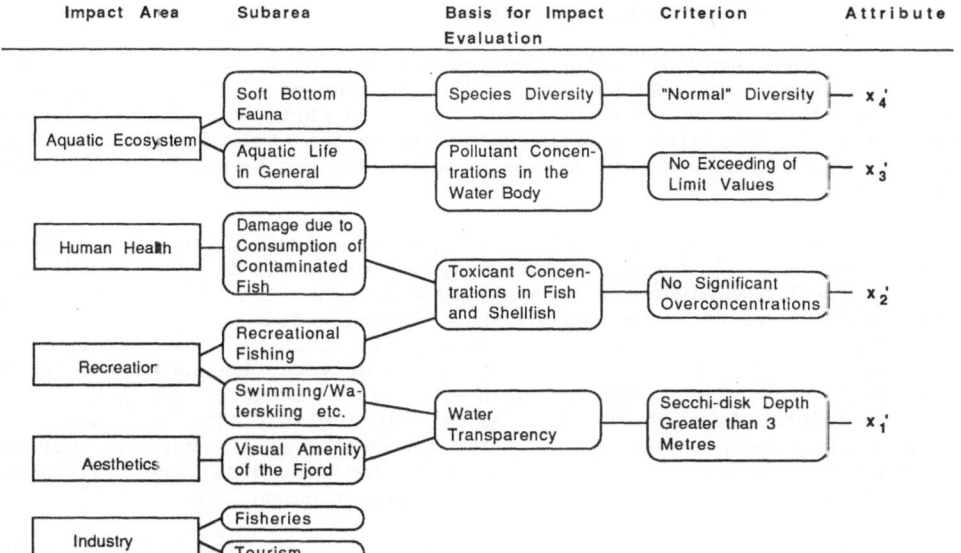

Impact Area	Subarea	Basis for Impact Evaluation	Criterion	Attribute

Figure 10.3   Derivation of a simplified set of attributes for the Kristiansand fjord. Each attribute x_1' - x_4' is defined as the area (in hectares) of the part of the fjord where the relevant criterion is not met

Similarly, the damage to recreational fishing and the risk to human health resulting from consumption of contaminated fish, both depend on how much toxicants are present in the edible fish species that stay in the fjord. It seems reasonable, therefore, to use content of toxicants in fish as an indicator of the impact on both of these areas [albeit the latter does not appear to be a big problem (Heiberg et al., 1987)]. The corresponding attribute (x_2') is defined as the area of that part of the Kristiansand fjord where fish contain toxicant concentrations at levels that could mean a threat to human health. No attempt has been made here, however, to define specific limit values for all relevant toxicants.

In the ecological area two damage measures have been defined. To assess the effects of pollution on aquatic life in general, pollutant concentrations in the water column were compared with available water quality guidelines for marine life to see in what areas damage is likely to occur. In addition to this, special consideration was given to the soft bottom fauna. A separate attribute was developed for this section of the ecosystem on the basis of the concept of species diversity (Hurlbert, 1971) and the establishment of a normal species diversity level (Rygg, 1984a). The final attributes $(x_3'$ and $x_4')$ were defined in analogy with the two previous ones, x_1' and x_2'.

In this way the original set of ten (unspecified) attributes has been reduced to a total of four relatively simple effect parameters. Further details concerning the application of these parameters are given in sections 5 and 7.

It appears evident that pollution control in the Kristiansand fjord should aim at reducing (1) the amount of dissolved and particulate organic matter in the inner part of the fjord, in particular in Vesterhavn-Fiskåbukta, (2) the level of organochlorines in the whole fjord area, and (3) the concentrations of metals in water and sediments in Vesterhavn-Fiskåbukta.

The local authorities in Kristiansand have drawn up extensive plans for reducing the discharge of municipal sewage into the fjord. In contrast, definite plans for controlling pollutant loadings from industry or the seabed have yet to be established. In particular, no specific plans currently exist for further reducing the discharge of metals and organochlorine compounds from Falconbridge Nikkelverk, the major source of these pollutants. The organochlorines pose a special problem. They originate unintentionally from the use of chlorine in different parts of the manufacturing process. The relative importance of the various sources is difficult to assess, and at present no obvious measures are available that could prevent the formation of the pertinent compounds.

Due to the paucity of definite plans, we have been forced to look at some hypothetical measures to control industrial pollution. In total, we have considered actions that will reduce the loading of (i) municipal sewage to Vesterhavn-Fiskåbukta, (ii) organic micropollutants to the whole fjord area, and (iii) heavy metals from Falconbridge Nikkelverk or from bottom sediments in Vesterhavn. A more detailed description of the actions investigated are given below.

Table 1

Action	Description	Estimated costs in Norwegian Kroner[1,2] (in millions)
A	Diverting the municipal sewage from the western part of Kristiansand to Vestergapet (see Figure 10.1) and building a sewage treatment plant	55
B	Effecting a full stop in the input of organic micropollutants to the fjord	no estimate available
C	Reducing the present discharge of nickel, copper, and iron from Falconbridge by 80%	100
D	Capping the seabed in most of Vesterhavn (ca. 2.3 km^2) using sand, silt, or clay as capping material	55-100

[1] All estimates are assumed to be in 1986 prices.
[2] Divide by 7 to roughly convert the amounts into mill. U.S. dollars.

The selection of metals under Action C was based on the results of a screening using the so-called Monitoring Trigger Level (MTL)-system in combination with data on the current discharge of metals from Falconbridge. The MTL-system, which was developed by the U.S. Environmental Protection Agency (Kingsbury and Chessin, 1984), constitutes a set of water quality guidelines for metals and various other chemical substances. An 80% reduction in the metal discharge has been suggested as an appropriate long-term goal by the Norwegian State Pollution Control Authority. Since no practical solution of how to bring about such a reduction has as yet been devised, the cost estimate of 100 million Norwegian Kroner (NOK) should be regarded as highly uncertain. It is meant to be an upper bound to the cost rather than a real estimate.

Action B is definitely the most hypothetical one since no thorough survey has as yet been made of the sources of organic micropollutants and since no specific measures have been proposed to combat these pollutants. Therefore no cost estimate has been assigned to this action.

5   EFFECTS OF CONTROL ACTIONS

Referring to Figure 1.2, it is clear that a key step in analyzing the above measures is to estimate how each of them will affect the attributes x_1' through x_4'. Water clarity in Vesterhavn-Fiskåbukta is affected largely by a single kind of pollution, municipal sewage. The damage to recreational fishing/human health is due mainly to the organochlorines that are present in some edible fish species. The levels of heavy metals found in the fish are not believed to pose a significant risk to humans. The soft bottom fauna, on the other hand, is affected primarily by heavy metal contamination. In this case, organics are probably playing only a minor role.

Because of the points just mentioned, in estimating the effects of pollution control we have made the simplifying assumption that Action A (municipal sewage) has an impact only on attribute x_1' (water clarity), that Action B (organic micropollutants) will affect attribute x_2' (fish toxicants) only, and that Action C (heavy metals) has a bearing on the ecological attributes x_3' and x_4' and on none of the other variables. Based on these assumptions, the effects of each action A through D are discussed in greater detail in the following sections.

5.1 Effect of Action A

As already mentioned water clarity is quantified in terms of the secchi-disk depth. The mean secchi-disk depths measured in Vesterhavn and Fiskåbukta are 7.1 and 6.8 metres, respectively (Molvær et al., 1986). Using the water quality criterion defined in section 3, these values indicate that satisfactory conditions for bathing in Vesterhavn-Fiskåbukta have already been obtained. Shorter periods of low water transparency do, however, occasionally occur during the summer season (a secchi-disk depth as low as 2.5 metres has been measured in Vesterhavn). Also, near Falconbridge Nikkelverk and at several other places close to the shore the water is generally quite turbid. Therefore, implementing Action A should be expected to provide a significant improvement in the water quality in Vesterhavn-Fiskåbukta.

Approximate secchi-disk depths can be established theoretically from concentrations of total phosphorus and nitrogen in the water column via calculations of chlorophyll $\underline{a}$ concentrations. However, with Action A the discharge of phosphorous and nitrogenous substances to Vesterhavn-Fiskåbukta should be eliminated or at least reduced to a very low level. Therefore, no detailed calculations seem necessary in this particular case. One can simply assume that the value of x_1' will be approximately zero after implementing Action A ($x_1'(A) \approx 0$).

5.2 Effect of Action B

As mentioned in section 2, people have been advised not to eat fish caught in an area of the fjord covering about 20 to 25 km². Since 1982 the discharge of organic micropollutants from the industry has been markedly reduced. Nevertheless, the level of organochlorines in fish and shellfish is still (1985) quite high, in particular in Vesterhavn and Fiskåbukta. Moreover, while there has been a decrease in the pollution level in certain species, one has not been able to observe a clear time trend for fish and shellfish as a whole. Therefore, without further discharge reductions, one must expect that in large parts of the fjord the fish will continue to contain unacceptably high concentrations of organochlorines.

With Action B, toxicant concentrations will gradually decrease and eventually reach a harmless level in the whole fjord area. Consequently, x_2^1 will attain a value close to zero after implementing Action B ($\bar{x}_2^1(B) \approx 0$). It is appropriate to point out, however, that unless special measures are taken (see Action D below), it will take a long time, probably several decades, before this result is obtained.

5.3 Effect of Action C

As will be recalled, Action C is to be assessed in terms of its impact on the concentrations in the water body of the relevant metals and on the diversity of the benthic community. Rygg (1984b) found a fairly high negative correlation ($r^2 = 0.62$, $n=59$) between the benthic community diversity and copper concentrations in sediments from a dozen Norwegian fjords, including the Kristiansand fjord. The regression equation derived was employed in the present work in combination with a general classification system for species diversity, also developed by Rygg (1984a). The system distinguishes between five different degrees of diversity: very low, low, moderate, normal, and high.

With this approach the central task in estimating the effect of Action C becomes that of predicting how that action affects metal concentrations (nickel, copper, iron) in various parts of the fjord, including the benthic sediments. To solve this task we have employed the Exposure Analysis Modeling System (EXAMS). This is a chemical fate and transport model for the aquatic environment developed by the U.S. Environmental Protection Agency (Burns et al., 1982). In applying EXAMS attention was confined to Vesterhavn-Fiskåbukta. This area was divided into three parts, inner Vesterhavn, outer Vesterhavn, and Fiskåbukta, the division being indicated by dotted lines in Figure 10.1. The volume below each subarea was split horizontally into three layers, a top water layer 4 metres deep, a bottom water layer (of variable depth), and a sediment layer 10 cm thick. Adaption of the model to the water regime in the area studied was done on the basis of nickel concentrations measured in the water column and the sediments. After calibration EXAMS was used to calculate steady-state concentrations of the relevant metals in the different segments and to simulate the time-development of the concentrations following a discharge reduction. A detailed account of these calculations can be found in Heiberg et al. (1987).

5.3.1 Water Concentrations. The maximum calculated steady-state concentrations of heavy metals in the water are shown in Table 10.1 along with corresponding water quality criteria for long-term exposure.

Table 10.1   Maximum Calculated Steady-State Concentrations in the Water Column and Corresponding AMTL[1] values. Concentrations in µg/l

	No action (present discharge)	Action C (80% reduction)	AMTL[1]
Ni	6.7	1.3	7.1
Cu	0.5	0.1	2
Fe	0.2	0.03	50

[1] Ambient Monitoring Trigger Level (Kingsbury and Chessin, 1984).

Assuming current discharge rates of the relevant metals, estimated steady-state concentrations in the water are all predicted to be below or at roughly the same level as the AMTL-values. This indicates that keeping the discharge from Falconbridge at the present level (the no-action alternative) would not mean a severe stress on aquatic life, even though some damage is likely to occur in the vicinity of the factory. According to the calculations, steady-state concentrations will be obtained in about 30 to 40 years in the bottom water layer (and in the sediments), whereas in the top water layer steady state is reached within a few weeks after a reduction in discharge.

With an 80% reduction of the discharge, estimated steady-state concentrations are 20% of those for the no-action alternative, indicating that Action C would give a high degree of protection to marine life in most of Vesterhavn-Fiskåbukta.

Unfortunately, the spatial resolution of the EXAMS calculations is much too coarse to allow a reliable prediction of the change in attribute x'_3 to be expected from Action C. It appears from the above results, however, that x'_3 is likely to be small both with Action C and under the no-action alternative.

5.3.2 Species Diversity. Table 10.2 shows the changes in species diversity of the soft bottom fauna expected to result from Action C. The results presented were derived from copper concentrations in sediments using the approach outlined above.

Table 10.2 indicates that if the discharge of copper is kept at the present level, species diversity will be below "normal" in all three parts

Table 10.2   Species Diversity of Soft Bottom Community in Vesterhavn-Fiskåbukta as Determined from Copper Concentrations in the Sediments

	Vesterhavn inner	Vesterhavn outer	Fiskåbukta
No action	low	moderate/normal	moderate
Action C	moderate	normal/high	normal

of Vesterhavn-Fiskåbukta and considerably lower in inner Vesterhavn. An 80% reduction of the discharge would lead to a significant improvement of the benthic diversity. However, even with such a reduction, diversity is still expected to be below "normal" in inner Vesterhavn.

As with attribute x_3', it is difficult to estimate what value of x_4' will result from Action C. A very rough estimate was made by using the information in Table 10.2 as a starting point and assuming, quite judgmentally, that the benthic fauna will be impaired in 50% of inner Vesterhavn, in 0% of outer Vesterhavn, and in 0% of Fiskåbukta. We have also assumed that without any further action the corresponding percentages will be 100, 25, and 50, respectively. In terms of areas this means that x_4'(no-action) $\approx$ 223 hectares, and x_4'(Action C) $\approx$ 65.5 hectares, both figures applying to the steady-state situation.

5.4 Effect of Action D

The purpose of sediment capping is to isolate pollutants in the sediments in order to prevent contamination of the overlying water body and protect organisms living on the seabed. Action D, capping of polluted sediments in Vesterhavn, will have a lasting effect only if carried out in conjunction with a substantial reduction in the discharge from industry, like that brought about by Actions B and C. Its main effect, in such a case, will be to reduce the time needed to achieve the full effect of the discharge reductions. Calculations carried out with EXAMS show that, in the case of Action C, metal steady-state concentrations will be reached up to 40 years earlier in the bottom water and benthic sediments in Vesterhavn and about 20 years earlier in Fiskåbukta (all segments).

In other words, sediment capping will have an influence on how rapidly the effects of Actions B or C, as described by attributes x_2' and $x_3' + x_4'$, respectively, are obtained, but will have no influence on the steady state itself. The time aspect is important, however, when it comes to valuation of the actions since effects obtained in the future are generally considered of less utility than effects obtained today. The discounting of utility will be further discussed in section 7 along with other details of Actions C and D.

6 VALUATION OF A TOTAL CLEANUP BY USE OF AN ECONOMIC METHOD

To establish an economic value of the benefits represented by a clean Kristiansand fjord, a poll was taken to survey people's opinions about this issue (Appendix E). The interviewees were given information on the current state of pollution in the fjord, on various abatement measures available, and on the effects to be expected from these measures. Specifically, the interviewees were told that after effecting the measures outlined, the water in the fjord would be as clean as that near other larger towns along the Norwegian coast and that fish caught in the fjord could again be eaten without any risk of health damage.

The main question was aimed at determining how much people would be willing to pay for obtaining a clean Kristiansand fjord. An individual's willingness-to-pay (WTP) was defined as the answer obtained to the following question:

"Certain provisions will be taken to reduce the pollution of the Kristiansand fjord, provided you and other citizens pay a certain amount of money each. What is the maximum amount you would be willing to pay for the above purpose?"

Two studies were conducted. For the first study 659 persons were selected randomly nationwide; for the second study an additional 300 persons were selected from the Kristiansand area. With a few exceptions the same set of questions were asked in both studies. In the local study, however, people were also asked about their WTP if abatement measures were to be financed through a permanent increase in the annual sewage excise tax and not as a single payment.

The national study was performed through personal interviews whereas in the local study questions were asked by telephone. Both surveys employed the so-called bidding-games technique. Further details can be obtained from the questionnaire, reproduced in Appendix E, and from the report of Heiberg et al. (1987). Although no attempt was made in the questionnaire to explain in detail what improvements would result from the measures under consideration, it is clear from the information offered that the revealed WTP can be taken to represent the value of a nearly total cleanup of the Kristiansand fjord.

6.1 Main Results

The main results from the WTP survey are summarized below.

6.1.1 Cne Fjord. For a single payment to clean up the Kristiansand fjord the average WTP per person was 511 NOK nationwide and 743 NOK locally. For a permanent annual increase in the sewage excise tax the average WTP locally was 331 NOK.

These results suggest that the Norwegian population is willing to contribute, as a single payment, about 1165 million NOK to clean up the Kristiansand fjord (2.28 million taxpayers each willing to spend 511 NOK on the average). Of this amount an estimated 22 million is derived from the citizens of Kristiansand. Moreover, the study indicates that this subgroup of the population is willing to contribute about 10 million NOK in annual sewage excise taxes.

6.1.2 Two Fjords. In both the national and local study people were also asked about their WTP for having the Frierfjord, another heavily polluted coastal area in Norway, cleaned up in addition to the Kristiansand fjord. The overall average WTP for these two fjords was found to be 577 NCK nationwide and 633 NOK locally.

The national result is only 10% higher than the WTP obtained for the Kristiansand fjord alone. This may be taken to mean that people's marginal willingness to pay for environmental benefits is rapidly decreasing, which is not very surprising. Among the local residents the WTP for two fjords is lower than for one fjord. This inconsistency is probably due to an unsatisfactory presentation of the question in the local study.

6.2 Conclusions

The aggregated WTP for cleaning up the Kristiansand fjord is more than a billion Norwegian Kroner. For such a sum one should be able to entirely solve the problem of pollution in the Kristiansand fjord. It thus appears that a total cleanup of the fjord would be cost-effective.

One cannot draw any definite conclusions from the WTP survey about the cost-effectiveness of individual measures with more limited environmental goals, such as those formulated in section 4. In the poll, questions were asked which aimed at providing a basis for evaluating limited measures. Specifically, the interviewees were presented with a list of environmental factors and asked to assess their relative importance by assigning a score to each one (see question XX1 in the questionnaire). The outcome of this part of the study was not very informative as people tended to give a high score to all factors. The two concerns that were rated highest, however, were "no risk by eating fish" and "no risk by swimming", i.e. both were related to health risk issues.

Apparently, more refined valuation methods are required if the utility of limited measures is to be assessed. In this work we have explored one formal method that could be usable for this purpose, the Simplified Multiattribute Rating Technique (SMART). Application of this technique is dealt with in the next section.

7   VALUATION OF POLLUTION CONTROL ACTIONS BY USE OF "SMART"

This section focuses on the modelling of people's preferences regarding pollution control in the Kristiansand fjord. The purpose of the modelling is to compare different proposals put forward to clean the fjord and to find out which proposals appear the most cost-effective.

7.1 Methodology

7.1.1 The SMART Approach. SMART (Edwards, 1977; von Winterfeldt and Edwards, 1986) can be viewed as a simplified version of the multiattribute utility analysis (MUA), a formal multi-objective valuation method developed by Keeney and Raiffa in the 1970s (Keeney and Raiffa, 1976) (see Chapter 1, section 3.5).

Multiattribute utility theory is based on a set of axioms concerning how people make decisions in the presence of uncertainty. A key point of the theory is the assumption that people's preferences can be described mathematically by a so-called utility function. People are assumed to act so as to maximize their utility, given the restrictions they meet, like limited income etc.

In SMART the utility function is assumed to be of the particularly simple additive form. A general expression for such a function is

$$U(\underline{x}) = \sum_{i}^{n} w_i u_i(x_i) \qquad \text{with} \qquad \sum_{i}^{n} w_i = 1, \qquad (10.1)$$

where $\underline{x} = x_1, x_2, \ldots, x_n$ is the vector of attributes used to describe the impact of the options under consideration, $U(\underline{x})$ is the overall utility of the situation represented by the vector $\underline{x}$, u_i is the utility function for the single attribute x_i, and w_i is the weight of that attribute.

In an ordinary MUA the weights w_i are determined through a fairly complicated interview process that may be demanding and time-consuming. The approach taken in SMART consists of estimating the weights directly through simple questions presented to the interviewee orally or in a questionnaire.

With regard to the adequacy of an additive model, Cox (1986, p. 129) argues that:

"In many, if not most, applications of multiattribute preference assessment, the additive form....for values or utilities has provided an adequate representation of individual preferences, and use of the more complicated multiplicative form has not been justified by a significant increase in predictive validity."

The special procedure used in this study, involving a postal survey among the general population in Kristiansand (section 7.2), allowed that only relatively simple questions could be asked. This also suggested the use of a utility function of a simple form.

7.1.2 Utility Function for the Kristiansand Fjord. The attributes used in the survey are specified below. They include three environmental variables, x_1, x_2, and x_3, derived from the attributes x_1', x_2', and x_4' defined in section 3.2, plus one cost variable, x_4.

- x_1 (water clarity) is the total area of the fjord where water transparency is sufficiently improved to allow swimming, bathing, and similar recreational activities, a secchi-disk depth of 3 metres being used as the criterion. The area is specified as <u>percentage</u> of the area currently not meeting the criterion. x_1 may vary between 0 and 100. A value of 0 corresponds to the present situation (or more correctly, to the situation eventually reached if the pollutant load is kept at the present level), whereas 100 corresponds to a total improvement of water clarity in Vesterhavn-Fiskåbukta.

- x_2 (recreational fishing/human health) is defined (intentionally somewhat loosely) as per cent change in the area of the fjord where fish contain significant amounts of potentially harmful chemicals. x_2 may range from 0 to 100.

- x_3 (soft bottom fauna) is defined as per cent change in the area of the fjord where species diversity of the soft bottom community is below an expected "normal" level. The range is from 0 to 100.

- x_4 (costs) is the total discounted cost of each countermeasure, in millions of Norwegian Kroner. In principle, this variable covers both investments and operational costs.

On the basis of the above attributes, a simplified utility hierarchy for the Kristiansand fjord can be constructed as shown in Figure 10.4. The figure suggests a two-step procedure for the assessment of people's preferences for a clean fjord and alternative goods (represented by the costs). First, the relevant environmental factors are evaluated and assigned weights (w_i) according to their relative importance. Next, a tradeoff is made between environmental improvements and the associated costs.

After the first step has been carried out, an environmental utility index U_e can be calculated for each measure as

$$U_e(x_1, x_2, x_3) = \sum_{i=1}^{3} w_i u_i(x_i) . \qquad (10.2)$$

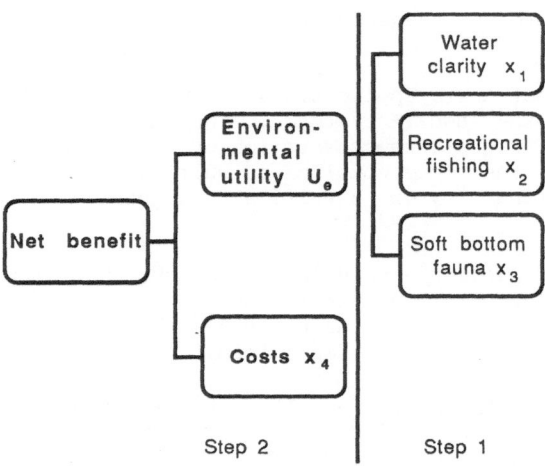

Figure 10.4   Simplified utility hierarchy for the Kristiansand fjord

By means of this index different measures can be compared and evaluated,
whatever their costs, in terms of how much they improve on the areas water
clarity, recreational fishing, and soft bottom fauna in the Kristiansand
fjord.

To employ equation (10.2), attribute values must have been established
and the form of each single-attribute utility function u_i must be known. In
this study we have assumed all the unidimensional utility functions to be
linear over the entire range of attribute values, 0 to 100. Although this
assumption may be questioned, it was considered impracticable to determine
the form of the functions u_i in greater detail through a postal interview
like that employed in the present study.

Following von Winterfeldt and Edwards (1986), we further scale the
single-attribute utilities so that u_i takes on values between 0 and 100.
This means that we have $u_i(x_i) = x_i$ (i=1,2,3).

In the second step of Figure 10.4, costs of countermeasures are taken
into account to yield a net benefit for environmental improvements in the
Kristiansand fjord. With a utility function of the form (10.1), one can
write the net benefit, B_{net}, as

$$B_{net} = k_e \, U_e(x_1, x_2, x_3) - x_4 , \qquad\qquad (10.3)$$

where k_e is a coefficient expressing the marginal rate of substitution
between money and a cleaner fjord. Equation (10.3) gives the net benefit,
in monetary terms, of an action leading to environmental improvements
described by the attribute values x_1, x_2, x_3, and costing x_4 million
Norwegian Kroner.

To establish the net benefit function, four coefficients must be
determined: w_1, w_2, w_3, and k_e. In this study the coefficients were
determined on the basis of a survey carried out among the general

population in Kristiansand. The details of this survey and the procedure used to obtain the required coefficients are dealt with in the next section.

7.2 The Survey

The survey was performed as a postal inquiry in the municipality of Kristiansand (population 61,000) during March 1987. A group of 190 people were randomly selected from the telephone book, stratified only by sex. After 4 weeks and one reminder, 70 answers were received. Of these, 63 are included in the calculations; the rest had to be discarded because of insufficient completion of the questionnaire.

7.2.1 The Questionnaire. The entire questionnaire is reprinted in Appendix F; only the main points will be discussed here.

First, a rather thorough description of the present state of pollution in the Kristiansand fjord is given, together with a map of the area of concern. Thereafter, three different effects of abatement measures are described, each representing a total improvement of one of the three environmental factors indicated in Figure 10.4.

The respondent is requested to evaluate these three effects in two steps. First (Question 1), the respondent is asked to rank the effects in decreasing order of importance. Second (Question 2), a value of 10 is assigned to the least important effect, and the respondent is asked to quantitatively rate the other two effects according to how much more important they appear to be by using one of the numbers 10, 20, 30, etc.

The purpose of Questions 3-5 is to find out how the respondent views the benefits of pollution control in relation to the economic costs involved. A specific abatement measure (Alternative I) is presented to him, described as a complete elimination of the sewage discharge to Vesterhavn and Fiskåbukta (effect A). This action will improve water transparency in that part of the fjord, but will have no significant impact on the other two effect areas considered. A cost estimate for this action is 55 million NOK. Divided among all households in Kristiansand, the cost is estimated to be about 300 NOK a year (implicitly for 10 years). The respondent is asked to decide whether the benefits of the action outweigh the costs (Question 3). If the answer is yes, costs are assigned a value of 10 and the respondent is asked to rate the benefits relative to the costs using a multiple of 10. Conversely, if the costs are considered too high in relation to the benefits obtained, benefits are given a value of 10 and costs are rated in a way similar to that described above (Questions 4-5).

In Questions 6-8 the same procedure is carried through once more, but with another control option (Alternative II). Under this alternative water clarity in Vesterhavn-Fiskåbukta becomes fully satisfactory (effect A), recreational fishing will become possible in the whole of the Kristiansand fjord (effect B), and the soft bottom fauna will gradually return to normal in most parts of Vesterhavn-Fiskåbukta (effect C). A cost estimate for this action is 200 million NOK.

In Question 9 the respondent is asked to assess alternatives I and II directly against each other.

Finally, a few background variables are included: sex, age, income, education, and occupation.

7.2.2 Processing of the Data. The responses to Questions 1 and 2 are used to derive the coefficients of the utility function U_e [equation (10.2)]. If a_i is the score assigned by the respondent to environmental factor no. i (i=1: water clarity, i=2: recreational fishing, i=3: soft bottom fauna), a corresponding normalized weight w_i can be calculated as

$$w_i = \frac{a_i}{\sum_{j=1}^{3} a_j} \qquad (i = 1,2,3). \qquad (10.4)$$

From Questions 3-5 and 6-8 estimates can be made of the coefficient k_e. The answers given to these questions tell how many times greater the respondent feels that the specified environmental benefits are relative to the associated costs, or how many times greater the respondent feels that the costs are relative to the benefits. Calling the ratio between benefits and costs r, we will take the respondent's answer to mean that he or she feels the effects achieved through the action considered to be exactly worth an expenditure of $r \cdot x_4$ million NOK, x_4 being the amount presented to the respondent as the real cost of the action. In terms of the utility function (10.3) this may be expressed as $B_{net}(x_1,x_2,x_3,r \cdot x_4) = k_e U_e - rx_4 = 0$, giving

$$k_e = \frac{r \cdot x_4}{U_e} . \qquad (10.5)$$

Since the respondent is asked to consider two different measures (Alternatives I and II), with different effects and different costs, two different estimates of k_e can be determined. If the assumptions made about the form of the respondent's utility function hold, and if he or she is capable of judging the two alternatives in a completely consistent manner, the two estimates should be equal.

To obtain the values of U_e relevant to Alternatives I and II, the effects of those alternatives must be described in terms of the attributes x_1, x_2, and x_3. Although only a qualitative description of the alternatives was given in the questionnaire, we will assume that the environmental scenario actually considered by the respondent in the case of Alternative I, is the one characterized by the values x_1=100, x_2=0, and x_3=0. Furthermore, for Alternative II the relevant attribute vector is taken to be (100,100,100). From equation (10.2) we then obtain $U_e = w_1 u_1(100) + w_2 u_2(0) + w_3 u_3(0) = 100 \cdot w_1$ for Alternative I, $U_e = w_1 u_1(100) + w_2 u_2(100) + w_3 u_3(100) = 100$ for Alternative II. Using these results along with the pertinent cost estimates we obtain from equation (10.5):

Alternative I: $\qquad k_e = r_I \cdot 55/(100 \cdot w_1)$, $\qquad\qquad$ (10.6a)

Alternative II: $\qquad k_e = r_{II} \cdot 200/100$. $\qquad\qquad$ (10.6b)

7.3 Results

7.3.1 Step 1: Weighting of Environmental Factors. Average environmental weights, obtained as the arithmetic mean of the individual answers, are shown in Figure 10.5.

In view of the outcome of the willingness-to-pay study (section 6.2), where health effects obtained the highest score, it is somewhat surprising that the soft bottom fauna scores that high (41%) in the postal survey. This result can partly be explained, however, by the fact that in the former survey health effects were explicitly mentioned, whereas in the present one they are only indirectly referred to (via fish contamination).

From the estimated weights, the following aggregate environmental utility function can be constructed for the Kristiansand fjord:

$$U_e(x_1, x_2, x_3) = 0.27x_1 + 0.32x_2 + 0.41x_3 . \tag{10.7}$$

7.3.2 Step 2: Balancing of Costs and Benefits. The frequency distributions of the responses to Questions 3-5 and 6-8 are shown in Figure 10.6.

For both alternatives the center of the distribution is to the right of the break-even point (one), indicating that most people judged the benefits as more important than the costs.

The geometric mean of the benefit-cost ratio r is 1.08 for Alternative I and 2.45 for Alternative II.

Aggregate values of the coefficient k_e were determined as the geometric mean of the k_e's obtained for each individual using equation (10.6a) or (10.6b). For Alternative I we get $k_e = 3.0$, for Alternative II, $k_e = 4.9$.

Finally, from the above results we construct the following aggregate net benefit functions for the Kristiansand fjord:

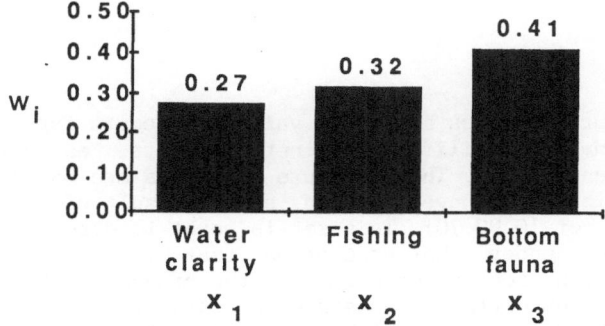

Figure 10.5 Normalized weights of environmental factors

$$\text{Alternative I:} \quad B_{net} = 3.0 \ (0.27x_1 + 0.32x_2 + 0.41x_3) - x_4 \quad (10.8a)$$

$$\text{Alternative II:} \quad B_{net} = 4.9 \ (0.27x_1 + 0.32x_2 + 0.41x_3) - x_4 \quad (10.8b)$$

In the next section these functions are used to evaluate specific countermeasures to the pollution of the fjord.

7.3.3 Valuation of Countermeasures. The countermeasures investigated are listed in Table 10.3. They include the four basic actions presented in section 4 (measures A through D) plus three different combinations of these actions (measures E through G). For completeness, the do-nothing alternative (measure H) is also included.

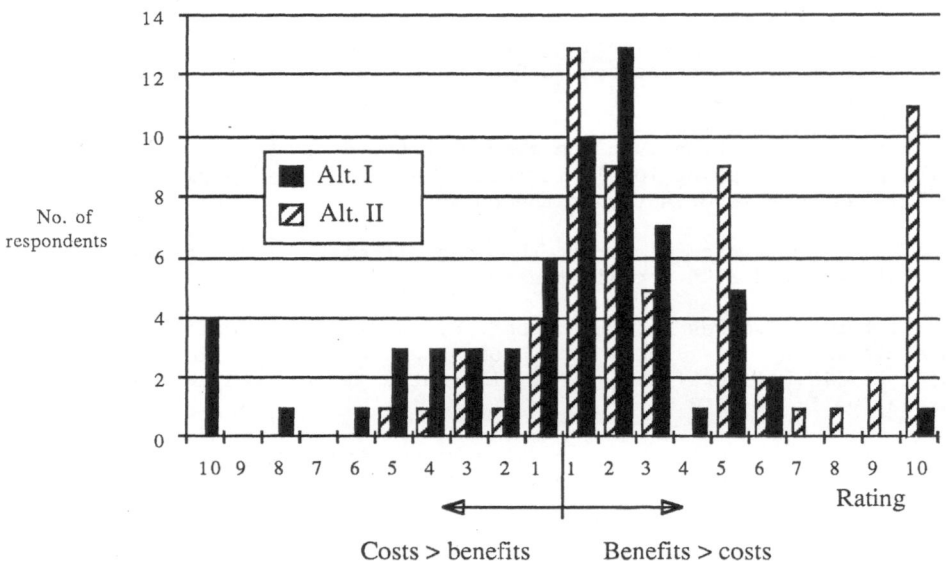

Figure 10.6   Distribution of responses for the cost/benefit (or benefit/cost) ratio

Table 10.3 further gives attribute values and costs for each measure, as well as environmental utilities and net benefits, as obtained from equations (10.7) and (10.8). The attribute estimates are based on the data presented in section 5. We have included, as the result of a "new" measure D', the attribute set (0,90,70). This set is meant to describe the effects of sediment capping under the assumption that the bed sediments in Vesterhavn are the present major source of the organic micropollutants. D and D' thus represent different possible effects of the same action: sediment capping in conjunction with a reduction of the metals discharge.

Table 10.3   Pollution Countermeasures in the Kristiansand Fjord

Measure	Clarity x_1	Fishing x_2	Fauna x_3	Environment U_e	Costs x_4	Net benefit Alt I	Net benefit Alt II
	Value				NOK mill.		
A: Sewage treatment[1]	100	0	0	27	55	27	78
B: Organic micropollut- ants	0	100	0	32	45	51	111
C: Heavy metals	0	0	70	29	100	-12	41
D: Seabed capping + C	0	0	70	29	200	-112	-59
D': " " "	0	90	70	57	200	-26	82
E: A+C	100	0	70	56	155	15	119
F: A+B+C[2]	100	100	70	88	200	66	230
G: A+B+D	100	100	70	88	300	-34	130
H: Status quo	0	0	0	0	0	0	0

[1] This is identical to Alternative I.
[2] This is almost Alternative II, except that people were told the impact would be (100, 100, 100).

The cost estimates presented correspond to those given in section 4, with the additional assumption that Action F, the most comprehensive measure not involving sediment capping, will cost 200 million NOK (compare with the cost estimate for Alternative II given in the questionnaire).

The fact that the full effect of the countermeasures are not achieved immediately is ignored in this section. The time aspect is relevant in connection with the measures involving sediment capping, i.e. D, D', and G, and will be discussed in section 7.3.4.

Figure 10.7 is a graphical presentation of the outcome of the analysis. Each point in the diagram corresponds to an action. The ordinate is the environmental index, as given by equation (10.7), the abscissa is the cost. The two straight lines passing through the origin are the graphs of the functions obtained by substituting B_{net} = 0 in equations (10.8a) and (10.8b). They determine the maximum amount people in Kristiansand would be willing to pay to obtain an improvement corresponding to a given environmental index (irrespective of the expenditure actually required), as judged from their assessments of Alternatives I and II, respectively.

The horizontal distance from a particular point in the diagram to one of the lines is the absolute value of the net benefit of the action described by the point concerned. The two lines roughly halve the plane spanned by the coordinate axes. The upper left part corresponds to actions with a positive net benefit, the lower right part to actions whose net benefit is negative.

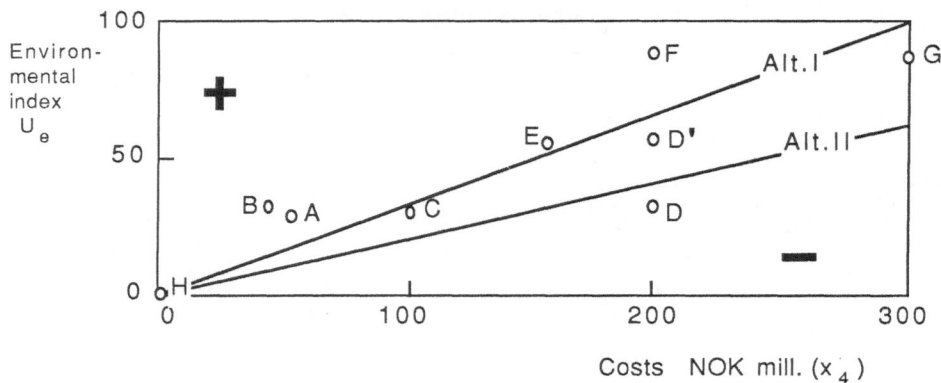

Figure 10.7 Cost-benefit diagram for pollution control in the
Kristiansand fjord

All eight investigated options are plotted in the benefit-cost diagram (Figure 10.7). As one can see, with Alternative II all but one option (measure D) are cost-effective (see also Table 10.3). Measures C, D', and G are cost-effective when Alternative II is used as the basis for the calculations, but they are not if Alternative I is chosen. The reason why measure D scores so poorly is that this measure includes sediment capping as a component. In the present <u>static</u> calculations, capping adds to the costs without contributing anything to the improvement of the environment. As indicated in section 5.4, the main effect of capping, as far as measures D and G are concerned, is to shorten the time needed to achieve the full effect of the other measures carried out in conjunction with the capping operation. The implications of this will be further discussed in section 7.3.4.

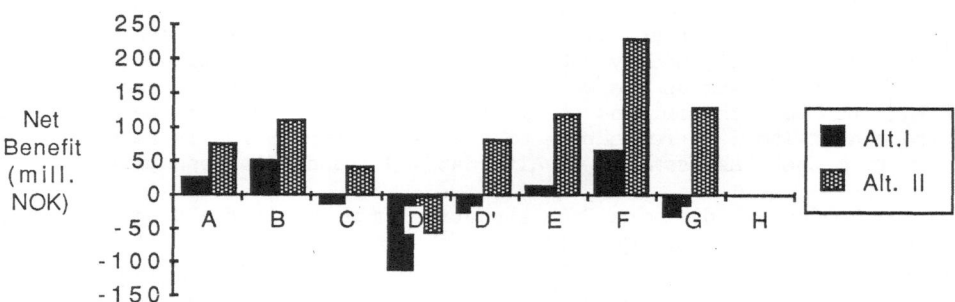

Figure 10.8 Estimated net benefits for measures A - H

The estimated net benefits are also displayed in Figure 10.8. The best option, according to the present calculations, is measure F. This conclusion holds with both net benefit functions. In the first case a net benefit to society of 66 million NOK is obtained, and in the second the net benefit is 230 million NOK.

Still another way of looking at the present results is to calculate, for each action, the net benefit <u>per NOK spent</u>. In Figure 10.9 the measures are listed in the order of decreasing net benefit to cost ratio. As can be seen the two cheapest measures, A and B, now become the most beneficial ones and, hence, are those that should be carried out first. One may notice that the order of the 8 measures is the same under both alternatives.

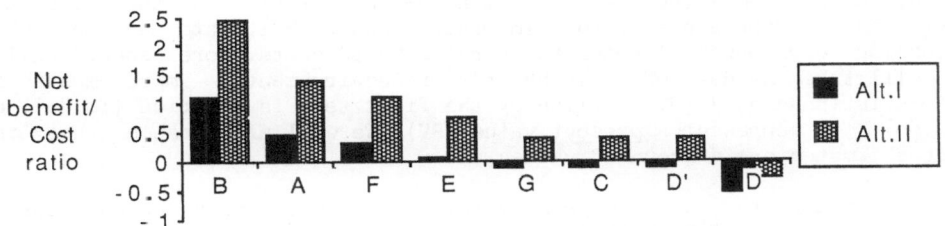

Figure 10.9 Net benefit/cost ratio for measures A - G

7.3.4 Sediment Capping. The analysis of countermeasures has so far been carried out as if benefits and costs occurred instantaneously. In reality, the effects will come gradually, the new steady state being reached perhaps after a long period of time.

In a normal investment analysis, future incomes and costs are discounted to a present value by means of a common discount rate. The formula normally used to obtain the present value (PV) of the (gross) benefit is

$$PV = \sum_{t=1}^{n} B_t \, (1+r)^{-t} \, , \qquad (10.9)$$

where B_t is the benefit of the project in the year $\underline{t}$, $\underline{r}$ is the discount rate, and $\underline{n}$ is the chosen time-horizon. An analogous equation is applied to obtain the present value of the costs. If the net present value of the project is positive with an appropriately chosen discount rate, the project is to be judged as economically favourable and should be put into effect.

Let us denote by B_0 the annual benefit of the project when its full effect has been achieved and by $f(t)$ the fraction of that benefit obtained in year $\underline{t}$. We will assume that the time-horizon is so long that the full effect is obtained before year $\underline{n}$. Equation (10.9) can then be rewritten as

$$PV = B_0 \cdot \sum_t^n f(t) \cdot (1+r)^{-t} \ . \tag{10.10}$$

In the present study Actions C and D (and F and G) both yield essentially the same final results but at different times (see the discussion in section 5.4). In other words, B_0 of the two actions are the same, but the time-development $f(t)$ differs. This implies that the present values of the two actions, as given by equation (10.10), will be different.

In the questionnaire, the dynamic aspect of the abatement problem was not very well accounted for. It was briefly mentioned that it would take a considerable time before the bottom fauna would be restored, while the time aspect was not referred to at all when dealing with diminished fish contamination. Nevertheless, we will assume here that those answering the questionnaire were aware of the time factor and took it into account when assessing environmental benefits in relation to costs. More specifically, we will take the derived gross benefit of environmental improvements in the Kristiansand fjord, as given by the first term in equation (10.8a) or (10.8b), to represent a present value (PV). We will also assume that the cost estimates are discounted values.

According to this the gross value of Action C, which one can derive from the data in Table 10.3, is to be considered a present value, $(PV)_C$. To determine the present value of Action D, $(PV)_D$, one can use equation (10.10) to obtain:

$$(PV)_D / (PV)_C = \sum_t^n f_D(t) \cdot (1+r)^{-t} \Big/ \sum_t^n f_C(t) \cdot (1+r)^{-t}, \tag{10.11}$$

where $f_C(t)$ and $f_D(t)$ are the time-development for Actions C and D, respectively. This expression shows that the present value of Action D (or G) can be determined by the following steps: estimating the functions $f(t)$ pertinent to Actions C and D (F and G), calculating the corresponding discount factors,

$$R = \sum_t^n f(t) \cdot (1+r)^{-t} \ , \tag{10.12}$$

and multiplying the estimated present value of Action C (F) by the ratio of the discount factors of Actions D and C (F and G).

The function $f(t)$, for a particular action, is given as $U_e(t)/U_e(n)$, where $U_e(t)$ is the environmental utility in year $\underline{t}$. This utility can be calculated from the corresponding attribute vector $\underline{x}(t)$ by means of equation (10.7). Consequently, to carry out an analysis like that outlined above, we need to know how the attribute values develop over time. Only in the case of Actions C and D do we have some information on this matter; thus, we will restrict ourselves in the following to this particular pair of actions.

The information available is that provided by the EXAMS calculations. Since these calculations give only a rough picture of the time-development of the pertinent attributes, the analysis of the dynamic aspects of Action D, detailed below, should be considered as illustrative, rather than as a highly realistic study. The procedure used to obtain $(PV)_D$ presupposes that the calculated time-development of Action C, $f_C(t)$, is identical to that envisaged by the respondent when making the trade-off between benefits and costs (of Alternative II). Since we cannot expect this to hold fully (basically because no detailed information on $f_C(t)$ was given to the respondent), there is a further reason not to take the results arrived at too literally.

The only relevant attribute in the case of Actions C and D is x_3 (x_1 and x_2 are constantly equal to zero). With regard to the time-development of this attribute, we have made the following assumptions:

Action C:	$x_3(0) = 0$,	$x_3(\infty) = 70$
Action D:	$x_3(0) = 45$,	$x_3(\infty) = 70$

The reason why sediment capping does not immediately give a full effect of 70% is the fact that only sediments in Vesterhavn are capped, whereas the total area considered also includes Fiskåbukta (see Table 10.2). From the results of the EXAMS calculations, the time-horizon was set to 40 years, meaning that steady state [i.e. $x_3(\infty)$] is effectively reached after 40 years. The same calculations indicate that a major part of the full effect is attained already after about 20 years. In the present calculations we have assumed that $x_3(20) = 0.80 \cdot x_3(\infty)$ for Action C, and $x_3(20) = 0.95 \cdot x_3(\infty)$ for Action D.

Assuming x_3 to develop exponentially with time, we obtain the functions shown in Figure 10.10.

Discount factors calculated from the above functions, applying different discount rates and a planning horizon of 40 years, are given in Table 10.4.

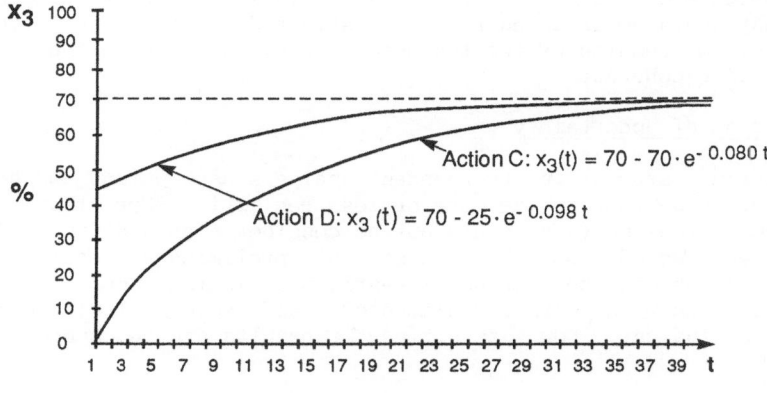

Figure 10.10   Time-development of attribute x_3 under Actions C and D

Table 10.4 Discount Factors (R) Obtained for Actions C and D

| Discount Rate | Action | |
	C	D
7%	4.94	7.94
4%	8.37	12.16
0%	19.93	25.62

In Table 10.3 the net benefit of Action D was given as -112 and -59 million NOK, under Alternative I and II, respectively. Using the figures in Table 10.4 corresponding corrected net present values may be calculated. If we adopt a discount rate of 7%, we see that the net present value of Action D can be obtained by multiplying the gross benefit ($k_e U_e$) by the fraction 7.94/4.94=1.61 and subtracting the cost. This gives a value of -61 million NOK in case of Alternative I and 26 million NOK under Alternative II. Under the second alternative the net benefit is positive. For Action C the corresponding net benefit was somewhat greater, 41 million NOK. This result suggests that the benefits of sediment capping (in terms of a quicker restoration of the bottom fauna) may not be big enough to justify the extra costs.

7.3.5 Alternative I versus Alternative II. In Question 9 people were requested to quantitatively rank Alternative I versus Alternative II. One may infer from the answers given to Questions 3-5 and 6-8 what alternative they should prefer, and also by how much. Question 9, therefore, provides redundant information, and was meant mainly as an extra check of the consistency in people's answers and of whether they had understood the preceding questions properly.

The geometric mean of the answers to Question 9 is 3.1 in favour of Alternative II, indicating that people judge Alternative II as considerably more preferable than Alternative I. Using equation (10.8a) the net benefits of Alternatives II and I can be calculated as 104 and 27 million NOK, respectively, which gives a ratio of 3.8. Similarly, using equation (10.8b), a value of 3.7 is obtained for the ratio of the benefits of Alternatives II and I. Thus, in a direct comparison of Alternatives I and II, people tend to be somewhat less strongly in favour of Alternative II than what could be expected from the estimated net benefits of the two alternatives. The results of the two assessments must, nevertheless, be said to be quite coherent.

7.4 Assessment of Uncertainty

At various stages in the present analysis assumptions and simplifications were made that may be more or less realistic. For this reason the outcome of the analysis should not be considered as a definite answer to the problem studied, but only as a set of preliminary results whose reliability needs to be further investigated. In this section the main types of uncertainty are briefly discussed, and some calculations are presented to indicate how sensitive the results are to changes in the assumptions.

First, uncertainty is connected with practically all the effect estimates. In particular, a 70% improvement in species diversity predicted to result from measures C, D, E, and G, although being based on the results of the EXAMS calculations, is no more than a rough estimate. If, as an example, we use 50% instead , we obtain the net benefits shown in Figure 10.11.

Comparing Figure 10.11 with Figure 10.8 gives an indication of how sensitive the outcome of the analysis is to the value of the attribute x_3. As can be seen, all alternatives with a nonzero value of x_3 (C,D,D',E,F,G) have now decreased net benefits. With Alternative II as the basis, Action F still appears to be the most beneficial. With Alternative I, Action B replaces Action F as the best option. It can be further seen that with Alternative I, the net benefit of Action E changes from positive to negative, and that the benefit of Action C becomes close to zero under Alternative II (the value is still positive).

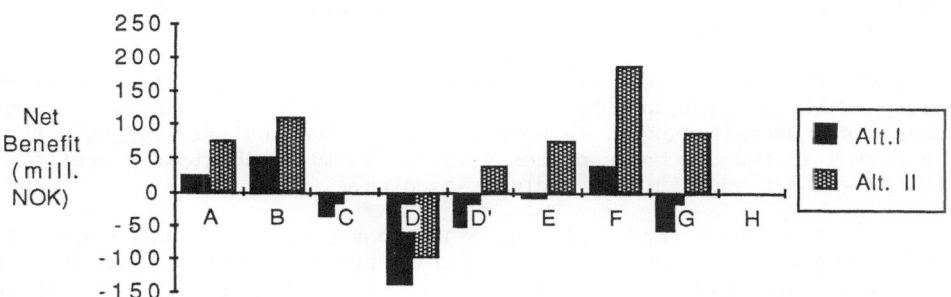

Figure 10.11   Sensitivity analysis. Net benefits obtained after changing x_3 to 50%

Second, the cost estimates involve considerable uncertainty. Apart from Action A all estimates employed should be considered as no more than informed guesses. As noted previously, even if costs are uncertain, one can obtain estimates from Figure 10.7 of the maximum willingness to pay for a specific action. Such an estimate provides an upper bound to how much the alternative may cost and still remain cost-effective. This kind of information may be very valuable to the decision-maker. As an example, according to Figure 10.7 implementation of Action B (a complete elimination of the input of organic micropollutants) would not be justified if the costs are appreciably higher than 170 million NOK.

Finally, in the valuation of measures, uncertainty is introduced both through the use of linear single-attribute utility functions and an additive preference model and through the coefficients entering the adopted model, i.e., w_1, w_2, w_3, and k_e. While moderate deviations from linearity are not likely to drastically alter the outcome of the analysis, the assumption about additive utilities might be more critical.

An illustration of the existing "model uncertainty" is provided by Figure 10.7, where, as was noted before, the two straight lines passing through the origin should coincide in an ideal case. The fact that they do not, may be taken to mean that people are not able to judge Alternatives I and II in an entirely consistent manner, resulting in an uncertainty in k_e. Alternatively, the discrepancy could mean that the form of the function used to model people's preferences is inadequate. It is interesting to note, though, that the difference between the two curves indicates an <u>increasing</u> marginal utility of environmental benefits, rather than a decreasing one, as would be expected. Irrespective of the cause of the difference, it is hardly possible to make any statements about the cost-effectiveness of actions corresponding to points within the wedge-shaped region bounded by the two lines.

7.5 Concluding Remarks

According to our analysis, the best control option is measure F. This is a combination of measures containing the following components:

- a full stop of sewage effluents to the inner parts of the fjord
- a full stop of the input of organic micropollutants
- an 80% reduction in the discharge of metals to the inner
  part of the fjord

With the given assumptions, this option will give a net benefit to society of 66-230 million NOK. The benefit of sediment capping, which brings in a dynamic aspect, is more uncertain. Our analysis suggests that a reduction in the discharge of heavy metals would be about as beneficial with and without a concurrent sediment capping.

The valuation method employed, SMART, has a number of advantages. It gives a simple and pedagogic description of the various conflicting objectives involved in the decision problem. The results are easy to understand both by experts and non-experts. Furthermore, the method can model the preferences of the general population. In principle, a large number of people can express their opinions. In general, this is not possible with other formal multiattribute methods as these methods normally involve lotteries and other difficult elicitation techniques.

One of the weaknesses of SMART is that the trade-off questions may be hard to answer. In particular, the questions requiring quantitative answers, i.e., the ratings, may be troublesome. This seems unavoidable, however, since these questions are at the very heart of the valuation problem.

The basic assumption in SMART is that people's preferences can be adequately described by an additive utility function. In this study we also assumed that all unidimensional utility functions were linear. While we have not gone to great length to test these assumptions, the results of the survey give some evidence that they may not be strictly valid. In particular, people's evaluations of Alternatives I and II do not exactly fit a preference model of the assumed form. It is possible, however, to interpret this finding as a result of people's inability to judge different alternatives in a completely consistent manner, rather than as an inadequacy of the preference model employed.

8 SUMMARY AND CONCLUSIONS

The present study describes two different ways of estimating the societal benefits of carrying out pollution abatement measures in the Kristiansand fjord. In the first survey the willingness-to-pay (WTP) for achieving a practically clean fjord was estimated through interviews carried out locally and among the entire Norwegian population. The result was a total WTP of 1165 million NOK nationwide and 22 million among the citizens of Kristiansand. Furthermore, the local residents would be willing to spend about 10 million NOK as an annual sewage excise tax. This corresponds to a present value of 133 million NOK, assuming a time-horizon of 10 years and a discount rate of 7%.

In the second survey, the citizens of Kristiansand were asked how they would valuate different abatement measures, leading to improvements within different parts of the environment. From their responses the benefits of various possible countermeasures were estimated using the Simplified Multiattribute Rating Technique (SMART). For the most comprehensive action considered (Action F), a gross benefit of 266-430 million NOK was obtained.

Action F is nearing a "full" cleanup of the Kristiansand fjord. Therefore, the corresponding benefit range, 266-430 million NOK, can be compared with the WTP obtained locally in the first part of the study. Compared with the annual and particularly with the once-and-for-all expenses, we see that the SMART survey indicates a substantially higher value of a clean Kristiansand fjord.

Both techniques employed in the present study provide useful means of valuating the benefits of pollution control. It seems clear, however, that of the two methods, SMART provides the more detailed picture of people's preferences for environmental goods, and is the method best suited for cases where the decision problem involves multiple or conflicting objectives.

9   ACKNOWLEDGEMENT

This study was funded by the Royal Norwegian Council for Scientific and Industrial Research. Thanks are due to Norman Green and Øivind Tryland of the Norwegian Institute for Water Research who greatly contributed to the study. Special thanks also to Marit Breivik and Ole Tollefsen for their valuable contributions to the willingness-to-pay and SMART surveys. The study has benefited from helpful and stimulating discussions with Professor James L. Regens during his period as a CCMS Fellow.

10   REFERENCES

Burns, L.A., Cline, D.M., and Lassiter, R.R., 1982, "Exposure Analysis Modeling System (EXAMS): User Manual and System Documentation," U.S. Environmental Protection Agency, Environmental Research Laboratory, Athens, Georgia, EPA-600/3-82-023.

Cox, L.A. Jr., 1986, Theory of regulaotory benefits assessment: econometric and expressed preference approaches, in: "Benefits Assessment: The State of the Art," J.D. Bentkover, V.T. Covello, and J. Mumpower, eds., Reidel, Dordrecht.

Edwards, W., 1977, Use of multiattribute measurement for social decision making, in: "Conflicting Objectives in Decisions," B.E. Bell, R.L. Keeney, and H. Raiffa, eds., John Wiley & Sons, New York.

Green, N.W., Knutzen, J., and Åsen, P.A., 1985, "Baseline Investigation of the Kristiansand Fjord. Report 3: Shallow Water Communities 1982-1983" (in Norwegian), Norwegian Institute for Water Research, Oslo, Report 8000354.

Heiberg, A., Hem, K.-G., Green, N., and Tryland, Ø., 1987, "Evaluation of Measures against the Pollution of the Kristiansand Fjord. An Example of a Cost-Benefit Analysis within the Environmental Sector," Center for Industrial Research, Oslo, Report 830123-5.

Hurlbert, S.N., 1971, The non-concept of species diversity, _Ecology_, 53:577.

Keeney, R.L., and Raiffa, H.F., 1976, "Decisions with Multiple Objectives: Preferences and Value Trade-offs," John Wiley & Sons, New York.

Kingsbury, G.L., and Chessin, R.L., 1984, "Monitoring Trigger Levels for Process Characterization Studies," U.S. Environmental Protection Agency, Industrial Environmental Research Laboratory, Research Triangle Park, North Carolina.

Knutzen, J., Martinsen, K., and Enger, B., 1986, "Baseline Investigation of the Kristiansand Fjord. Report 4: Contaminants in Fish and other Organisms" (in Norwegian), Norewgian Institute for Water Research, Oslo, Report 8000356.

Knutzen, J., and Martinsen, K., 1986, "Measure-orientated Monitoring of Micropollutants in Fish and other Organisms from the Kristiansand Fjord 1985" (in Norwegian), Norwegian Institute for Water Research, Oslo, Monitoring Report 262/86.

Molvær, J., Solheim, H.I., and Källqvist, T., 1986, "Baseline Investigation of the Kristiansand Fjord. Report 5: Water Replacement and Water Quality" (in Norwegian), Norwegian Institute for Water Research, Oslo, Report O-8000352.

Næs, K., 1985, "Baseline Investigation of the Kristiansand Fjord. Report 2: Metals in the Water Masses, Metals and Organic Pollutants in the Sediments, 1983" (in Norwegian), Norwegian Institute for Water Research, Oslo, Report 8000353.

Rygg, B., 1984a, "Soft Bottom Fauna Investigations. A Useful Tool in the Assessment of Marine Receiving Waters" (in Norwegian), Norwegian Institute for Water Research, Oslo, OF-80612.

Rygg, B., 1984b, Ecological detrimental effects of copper pollution in the marine environment, _Vann_, 4:464 (in Norwegian).

Rygg, B., 1985, "Baseline Investigation of the Kristiansand Fjord. Report 1: Soft Bottom Fauna Investigations 1983" (in Norwegian), Norwegian Institute for Water Research, Oslo, Report 8000355.

Tollefsen, O., Hem, K.-G., and Heiberg, A., 1988, "Evaluation of Pollution Countermeasures in the Kristiansand Fjord by use of the Simplified Multiattribute Rating Technique" (in Norwegian), Center for Industrial Research, Oslo, Report 830123-7.

Von Winterfeldt, D., and Edwards, W., 1986, "Decision Analysis and Behavioral Research," Cambridge, London.

ABATEMENT OF AIR POLLUTION IN OSLO

D. H. Trønnes

Center for Industrial Research
Oslo, Norway

1   INTRODUCTION

Some air pollution components in Oslo exceed the air quality guidelines proposed by the Norwegian State Pollution Control Authority (SFT, 1982). The situation is particularly adverse during the winter time on cold days with inversion and little wind. The incidence of lung diseases is considerably higher in Oslo than in other parts of Norway. The lung cancer incidence is 1.8 times higher in Oslo than in the rest of Norway (Aune, 1982), and air pollution is probably contributing to this together with variations in occupational exposure, smoking habits, diet, and other life style factors. The air pollution may also cause other health effects (Aune, 1982), and a survey has shown that about half of the population in Oslo feel afflicted by dust and unpleasant odours (Norsk Opinionsinstitutt, 1986). Prognoses for the situation towards the year 2000 show an aggravation with respect to some pollution components caused by an expected increase in road traffic and energy consumption, if no further actions are taken to reduce emissions. The already approved decision to require catalytic purification of exhaust in all new cars from 1989, however, will help to improve the situation for nitrogen oxides (NO_x), carbon monoxide (CO) and hydrocarbons.

To improve the air quality in Oslo the SFT has proposed a number of countermeasures which they want to consider for possible implementation (SFT, 1985). The number of possible abatement measures is great because there are many pollutants to be regulated and many different sources for the emissions. The resources to implement abatement measures are limited, so it is important to analyze all of the alternatives carefully.

The initiative to perform the present analysis was taken by the SFT. The goal of the analysis was to find out whether the benefit of an abatement measure outweighs the cost, to find which measures give the greatest benefit per unit costs, and to rank the measures according to their benefit/cost ratio. However, the motivations for performing the analysis are many: to gain insights about the abatement measures, to have a framework for the collection of information about the measures, to make more explicit assumptions, to make a clearer distinction between effect assessments and value judgments, to facilitate communication of the results to the decision-makers and to the public, to perform uncertainty and sensitivity analyses, and to document the results so they can be used to support political decisions about the implementation of abatement measures.

The present analysis has been organized and coordinated by the SFT.

2 METHODOLOGICAL APPROACH

In this analysis a multiattribute technique has been used. The technique is a simplified version of multiattribute utility theory (MAUT). For a detailed description of MAUT, see Keeney and Raiffa (1976). The procedure used in this study comprises six steps: characterizing the abatement measures, specifying the objectives, determining the impacts of the measures, valuating the desirability of the various impacts, calculating the benefit/cost ratios, and ranking the abatement measures.

In section 3 the abatement measures considered in this analysis are described. In section 4 the objectives of the abatement measures are formulated. The construction of an objectives hierarchy is very useful in structuring the objectives. The objectives are specified in the hierarchy until operational definitions, called attributes, are reached at the lowest level in the hierarchy. Section 5 discusses the impacts of the abatement measures in terms of the attributes. This step is often referred to as risk or impact assessment. In section 6 the benefit/cost ratios of the abatement measures are suggested as criteria for evaluating their goodness. To calculate the "benefit" it is necessary to perform a valuation of the effects in monetary terms or a weighting of the relative importance of the attributes. The weighting of attributes involves subjective judgements. On the basis of the impact assessments and the valuation of the effects, the benefit/cost ratios are calculated in section 7, and the countermeasures are ranked according to this ratio.

Often there are great uncertainties in the estimates from the risk and impact assessments. This is important information, because the decision-makers should know about the uncertainties in the outcome from implementing the countermeasures. In section 8 the uncertainties in the impacts and cost for one of the abatement measures have been assessed, and the resulting uncertainty in the outcome is calculated. The weighting of attributes is subjective and one can expect that the preferences will vary between individuals and in time. Section 9 presents the results of a sensitivity analysis that has been performed to study how variations in preferences affects the benefit/cost ratios and the ranking of the countermeasures. In section 10 some closing remarks are given.

3 COUNTERMEASURES

Road traffic and space heating are the most important sources of air pollution in Oslo, but industry and sewage incinerators also contribute significantly to the problem. Each of these sources will potentially give rise to at least one abatement measure. Further, there are many possible ways to reduce human exposure to the pollutants: reduction in the emissions by either reducing the activity level or by reducing the coefficients of emission or by relocating a given activity to a place where fewer people are affected. Combinations of different types of abatement measures with different sources give rise to a great number of possible measures to reduce the exposure to air pollution in Oslo.

A total of 32 different measures have been identified and considered in this study. These measures are grouped into 11 different categories:

1. Improved public transportation services.
2. Limitations on the use of private cars in the city center.
3. Regulations/controls of exhaust emissions from cars.
4. Reduction of dust in and around streets.
5. Alternative fuel for buses.
6. Hydroelectric power substitutes for fossil fuel.
7. Space heating (hot water based) from remote sources.
8. Energy saving measures.
9. Reduction/control of emissions from stationary sources.
10. Reduction of sulphur content in fuel oil and oil used for space heating.
11. Use of vegetation for absorption and as screens for air pollution/dust from road traffic.

The abatement measures in categories 1 through 5 and 11 are directed toward reducing air pollution caused by road traffic. Those in categories 6 through 10 are directed toward the heating of buildings, and some of the measures in category 9 toward emissions from industry. Among the road traffic measures, categories 1 and 2 can be viewed as a combined effort to reduce private motoring in the city center where the pollution concentrations are highest and many people are exposed. Categories 3 and 5 are aimed at reducing the coefficients of emissions, that is to reduce emissions per kilometer driven. Categories 4 and 11 are mainly meant to reduce the exposure to dust by putting a ban on the use of studded tires and to reduce exposure to dust by improved cleaning of the streets and by vegetation screens between streets and pavements (category 11). Among the space heating countermeasures some are directed toward using of less polluting energy sources (category 6), reduction and relocation of emissions (category 7), reduction in the demand for energy (category 8), and reduction in the use of oil types with high sulphur content. Within some categories many abatement measures have been defined. A complete list of all measures is given in Table 11.5 of section 7.

4 OBJECTIVES AND ATTRIBUTES

Before collecting the data about the abatement measures and carrying out the analysis, it is important to bound the problem and to structure the objectives. The overall objective of the abatement of air pollution in Oslo is to find the best pollution control action. The primary motivation for the considered abatement measures is to reduce human exposure to air pollution. However, to get a complete picture, all relevant impacts of the countermeasures and the costs must be considered. In total, five general concerns have been identified: exposure, well-being, acidification, side effects, and costs, as shown in the second level of the objectives hierarchy in Figure 11.1. The meaning of each general concern is defined by the objectives given below it in the hierarchy. The objectives listed in the objectives hierarchy also represent the bounding of the problem. Concerns that are not included here will not be taken into account in the formal analysis. Acceptance by the decision-makers and the involved interest groups of the objectives hierarchy is a prerequisite for acceptance of the results of the analysis. A brief description of the five general concerns considered in this analysis follows.

Reduce Exposure. As mentioned in the introduction, exposure to air pollutants can contribute to long-term health effects such as lung cancer and other lung and heart diseases. An obvious goal for measures aimed at improving air quality is, therefore, to reduce harmful health effects. In the first version of the objectives hierarchy "reduce health effects" was one of the five general concerns. However, to relate exposure to health effects, dose-response models are needed. Because models that can accurately predict the dose-response relationships are not available, it was considered to be too speculative and controversial to quantify health risks. Therefore the impact assessment had to stop at the exposure assessment, and the objectives hierarchy had to be changed accordingly. In particular, the objective "reduce exposure" replaced "reduce health effects." The objective "reduce exposure" is decomposed into reducing the number of persons exposed to sulfur dioxide (SO_2), suspended particulate matter (SPM), NO_x, and CO in concentrations above established guidelines. The guidelines used are those suggested by the SFT (1982). They are 100 $\mu g/m^3$ with 24 hours averaging time for both SO_2, SPM, and NO_x and 10 mg/m^3 with 8 hours averaging time for CO. In the analysis it is implicitly assumed that exposure to concentrations above the guidelines can contribute to long-term health effects and therefore should be reduced. Since quantification of health effects has been omitted, the responsibility of assessing the severity of exposure is left to the decision-makers in the sense that they will have to make value trade-offs between exposure and other objectives. This methodological problem will be discussed later in section 9.

Increase Well-being. As mentioned in the introduction about 50% of the population in Oslo report that they are afflicted by dust and unpleasant odour. It is therefore an important objective to increase people´s well-being by reducing the number of persons afflicted.

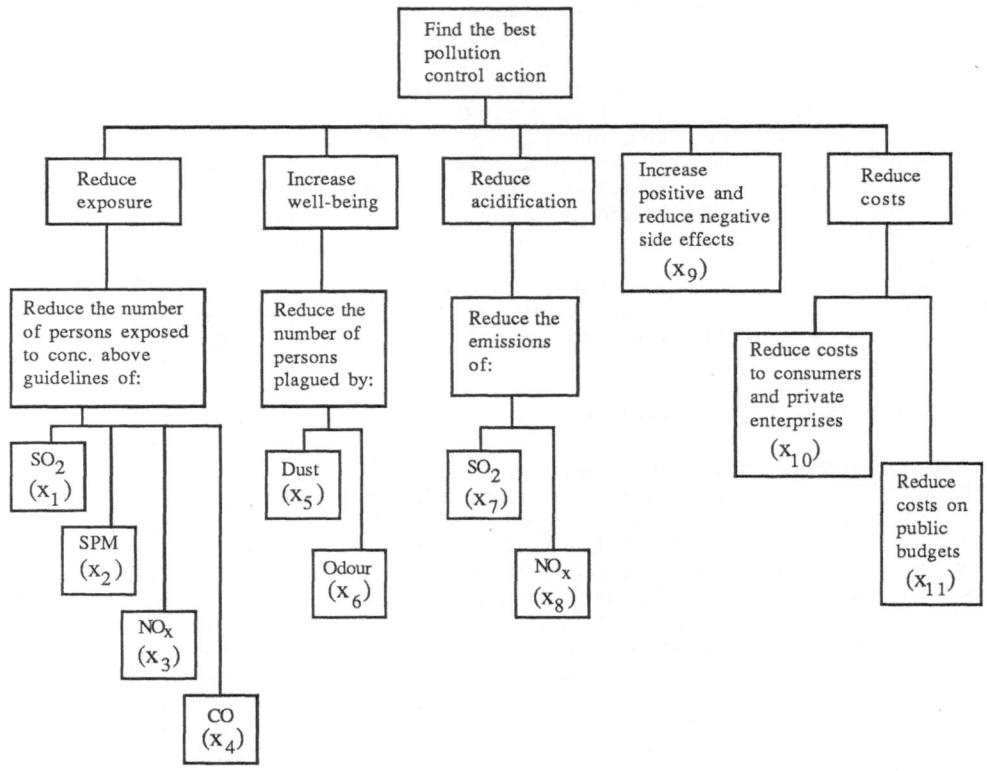

Figure 11.1   Objectives hierarchy for the abatement of air pollution in Oslo

Reduce Acidification.   Thousands of lakes in southern Norway have lost their fish populations as a result of acidification (see Overrein et al., 1980). There is also concern about acidification of the soil and effects on vegetation. Emissions of SO_2 and NO_x in Oslo contribute to these damages, and it is therefore desirable to reduce these emissions.

Side Effects.   This objective is included in the objectives hierarchy to account for any positive or negative side effects that may result from the countermeasures other than those explicitly mentioned in the hierarchy. Examples of side effects are: changes in travel times, changes in the number of traffic accidents, changes in the number of persons afflicted by noise, changes in expenses to heating, etc. The side effects are directly converted into monetary terms.

Costs.   Implementation of the countermeasures necessarily will result in costs. Some investments and running expenses will be covered by public budgets and some will be covered by the private sector. Naturally, reducing the costs of implementing the countermeasures is a very important objective.

Variables or attributes are assigned to each of the lowest-level objectives. By means of these attributes the impact of an abatement measure can be described quantitatively. The attributes employed are shown in Table 11.1.

Table 11.1 Attributes for the Abatement of Air Pollution in Oslo [1]

Objectives	Attributes	Units
Exposure		
1. Reduce the number of persons exposed to SO_2-conc. above guidelines	x_1 = Number of persons exposed to SO_2-conc. above guidelines	Persons (exposed to SO_2)
2. Reduce the number of persons exposed to SPM-conc. above guidelines	x_2 = Number of persons exposed to SPM-conc. above guidelines	Persons (exposed to SPM)
3. Reduce the number of persons exposed to NO_x-conc. above guidelines	x_3 = Number of persons exposed to NO_x-conc. above guidelines	Persons (exposed to NO_x)
4. Reduce the number of persons exposed to CO-conc. above guidelines	x_4 = Number of persons exposed to CO-conc. above guidelines	Persons (exposed to CO)
Well-being		
5. Reduce the number of persons afflicted by dust	x_5 = Number of persons afflicted by dust	Persons (afflicted by dust)
6. Reduce the number of persons afflicted by odour	x_6 = Number of persons afflicted by odour	Persons (afflicted by odour)
Acidification		
7. Reduce the emissions of SO_2	x_7 = Emissions of SO_2	Tons of SO_2 emitted/year
8. Reduce the emissions of NO_x	x_8 = Emissions of NO_x	Tons of NO_x emitted/year
Side effects		
9. Reduce negative and increase positive side effetcts	x_9 = Annualized present value of net change in side effects	10^6 NOK/year [2] (1987 prices)
Costs		
10. Reduce costs for consumers and private enterprises	x_{10} = Annualized present value of costs for consumers and private enterprises	10^6 NOK/year [2] (1987 prices)
11. Reduce costs on public budgets	x_{11} = Annualized present value of costs on public budgets	10^6 NOK/year [2] (1987 prices)

[1] By "persons exposed" in this table we mean exposure to concentrations above the guidelines
[2] 1 NOK = $ 0.15

In section 5 the impact of the countermeasures will be determined in terms of these attributes. Next, in section 6 the attributes will be valued in monetary terms, which will make it possible to compare the importance of the various benefit attributes and the costs.

5   ESTIMATING THE IMPACT OF THE COUNTERMEASURES

5.1   The Reference State

To measure the impact of the abatement measures one has to compare with some reference state. Here, the scenario to be expected, if none of the abatement measures considered in this analysis are implemented, defines the reference state. The reference state can be denoted by the vector of attribute values, x^0:

$$x^0 = (x^0_1, \ x^0_2, \ldots \ldots, x^0_8, \ 0, \ 0, \ 0). \tag{11.1}$$

The last three attributes will take the value 0, because there are no side effects or costs involved if no abatement measures are implemented.

The reference scenario is not fully defined before deciding on a specific year because of expected changes in energy consumption, road traffic, etc. Also, some abatement measures are already approved for implementation and in the future will affect pollution independently of the measures considered here. The installation of catalytic converters in all new cars beginning in 1989 is an example of a measure that will be implemented and will contribute to considerable reductions in exposure to NO_x and CO. The effect of this abatement measure will increase gradually as the fleet of cars is replaced, and it is expected to reach its full effect by the year 2000. It is therefore convenient to define the reference state to be the pollution situation in the year 2000 if none of the measures considered in this analysis are implemented.

The values of the attributes for the reference state are shown in the last row of Table 11.2. From the table one can see that in the year 2000, 184,000 and 150,000 persons will be exposed to concentrations above the guidelines for SO_2 and SPM, respectively. On the other hand only 12,000 and 0 persons will be exposed to concentrations above the guidelines for NO_x and CO, respectively, which, to a large extent, is a result of the decision to install catalytic converters in new cars beginning in 1989.

5.2   Exposure Assessment and Impact Assessment

The aim of the impact assessment is to quantify the effects of the abatement measures in terms of the attributes or performance measures defined in Table 11.1. The impact of abatement measure i, can be characterized by the vector of attribute values, x^i, describing the situation in the year 2000 if measure i has been carried out:

$$x^i = (x^i_1, \ x^i_2, \ldots \ldots, x^i_8, \ x^i_9, \ x^i_{10}, \ x^i_{11}). \tag{11.2}$$

Often it is more convenient to describe the impact of an abatement measure in terms of improvements in the performance measures together with its side effects and costs. This results in the vector x^{i*}:

$$
\begin{aligned}
x^{i*} &= (x^{i*}_1, \ x^{i*}_2, \ldots \ldots, x^{i*}_8, \ x^i_9, \ x^i_{10}, \ x^i_{11}) \\
&= (x^0_1 - x^i_1, \ x^0_2 - x^i_2, \ldots \ldots, x^0_8 - x^i_8, \ x^i_9, \ x^i_{10}, \ x^i_{11}).
\end{aligned} \tag{11.3}
$$

In the rest of this chapter the same vector will be used, but the superscript $*$ will be dropped for convenience.

The first step in preparing an impact assessment involves detailed specifications of each abatement measure. Then costs, reduction in emissions, and side effects are estimated. Most of this work has been carried out by institutions and organizations working in the respective fields.

Table 11.2 Improvements of the Attributes[1], x_1 - x_8, for the Abatement Measures and Attribute Values for the Pollution Situation in 1985 and in the Year 2000 (Reference State)

Abatement measure	Attributes SO$_2$ (x_1)	SPM (x_2)	NO$_x$ (x_3)	CO (x_4)	Dust (x_5)	Odour (x_6)	SO$_2$ (x_7)	NO$_x$ (x_8)	Correction factor [2]
1&2	2630	14630	9000	0	40000	7500	23	500	0.8
3.2	0	19900	0	0	3500	18000	0	0	1
3.3	0	0	0	0	0	630	0	0	1
3.4 [3]	0	7024	410	0	1619	3930	0	138	0.84
3.5 [3]	0	45000	15000	0	14050	100000	0	7000	0.84
3.6 [3]	0	32020	2020	0	8390	62500	0	768	0.84
4.1a [3]	0	0	0	0	800000	0	0	0	1
4.1b [3]	0	0	0	0	300000	0	0	0	1
4.2	0	0	0	0	23500	0	0	0	1
5.1	330	1500	630	0	300	1300	4	44	0.8
5.2	0	2750	1080	0	480	2380	0	100	0.8
5.3	0	900	0	0	160	780	0	0	1
5.4	2250	3330	1080	0	600	2880	18	108	0.8
5.5	2760	390	1290	0	70	360	2	120	0.76
5.6	1950	3100	0	0	550	2760	16	0	0.37
6.1	0	2640	0	0	650	540	0	0	0.92
6.2	3010	240	0	0	0	0	31	0	0.83
7D	109750	10500	0	0	0	0	-116	0	0.8
7F	109750	10500	0	0	0	0	756	0	0.8
7I	109890	10500	0	0	0	0	-87	0	0.8
8	2440	0	0	0	0	0	31	0	0.82
9.1	0	40810	0	0	7700	5540	0	0	0.74
9.2	16130	1750	0	0	2750	1250	200	0	0.8
9.3	0	0	0	0	22530	25060	0	0	0.87
9.4	0	500	0	0	9800	6200	0	0	1
10.1a [3]	180000	0	0	0	0	0	16780	0	1
10.1b [3]	25000	0	0	0	0	0	3000	0	1
10.3	39400	0	0	0	0	0	460	0	1
10.4	12900	6400	0	0	3000	7000	250	0	1
10.6	36900	6600	0	0	200	900	65	17	1
10.7	47100	6500	0	0	180	960	380	11	1
11	0	0	0	0	46880	0	0	0	0.8
Situation in 1985	30000	130000	100000	1000	248000	216000	2700	8600	
Reference state	184000	150000	12000	0	306000	230000	5160	7800	

[1] See Table 11.1 for definitions of the attributes
[2] See section 5.3 for explanation
[3] The abatement measure will be implemented on a national basis, and the improvements may therefore be greater than the attribute values for the reference state which refers to the situation in Oslo only

Table 11.3   Side Effects and Costs of the Abatement Measures

Abatement measure	Attributes[1]			
	Side effects (x_9)	Private costs (x_{10})	Public costs (x_{11})	SUM costs $(x_{10} + x_{11})$
1&2	750	89	162	251
3.2	1.5	4	1	5
3.3	0	0	0.022	0.022
3.4	0	2.8	0	2.8
3.5	1.26	131	0	131
3.6	0.13	30	0	30
4.1a	473	34	500	534
4.1b	189	10	150	160
4.2	0	0	2	2
5.1	1.04	0	1.6	1.6
5.2	0	0	0.41	0.41
5.3	0	0	0.87	0.87
5.4	0	0	20	20
5.5	0	0	2	2
5.6	0	0	12	12
6.1	1.38	0.6	2.9	3.5
6.2	0.2	2.9	0	2.9
7D	227	0	308	308
7F	227	0	312	312
7I	497	0	673	673
8	7.4	3.1	1.2	4.3
9.1	0	22.8	0.2	23
9.2	46	20	0	20
9.3	0	7.5	0	7.5
9.4	0.6	0	1.36	1.36
10.1a	55	101	0	101
10.1b	9	20	0	20
10.3	3	23	0	23
10.4	1.7	35	0	35
10.6	0.4	41	0	41
10.7	0.4	7.1	0	7.1
11	0	0	20.4	20.4

[1] Attribute values are given in 10^6 NOK   (1 NOK = $ 0.15)

For example, the Institute of Transport Economics (TØI), the Oslo Transportation Company, and the municipality of Oslo have been working with the traffic regulating measures, and Oslo Electricity Works has been working with some of the measures related to space heating.

Performing the exposure assessment is a very comprehensive task. The goal of this assessment is to obtain estimates of both the number of people that are exposed to various concentrations of the pollution components of interest, as well as the reduced exposure resulting from implementation of the abatement measures. To find the geographical distribution of the pollutant concentrations, one has to model the dispersion of the pollutant. Dispersion models contain equations describing the fate of air pollution components under different meteorological conditions like wind direction and velocity, whether it is inversion or not, etc. Inputs to these models are the amount and location of emissions. The output can be obtained in terms of "concentration maps" which give the concentration of each component for every square kilometer cell on a map. In Oslo, road traffic is an important source of air pollution, so the highest concentrations often occur on streets with heavy traffic. The Norwegian Institute for Air Research (NILU) has developed dispersion models that describe dispersion in streets and more global models for Oslo (see Grønskei et al., 1982). Combined with data on residence, work place, and travelling habits of the population, it is possible to obtain estimates of how many people are exposed to various concentrations of the pollutants. In this study NILU performed most of the work on the exposure assessments.

The assessment of the number of persons that are afflicted by dust and unpleasant odours is based on a survey in which a representative sample of the population in Oslo was interviewed (see Norsk Opinionsinstitutt, 1986). In the interviews people were also asked which sources contributed most to their afflictions. The effects of the abatement measures were then estimated by the SFT on the basis of how much each measure would contribute to the reductions in the relevant sources in various parts of the city.

Obvious the estimated impacts contain great uncertainties. In particular, one would expect the uncertainties in the exposure and well-being attributes to be large since these assessments are based on many uncertain elements like amounts emitted, distribution of the pollutants, and distribution of the population at various times of the day. No comprehensive quantification of uncertainties has been carried out in this study. However, in section 8 we take a closer look at one of the measures and show how the uncertainties can be quantified by a subjective assessment, and further how these uncertainties can be handled and incorporated in the analysis. The results of the impact assessments are given in Tables 11.2 and 11.3.

5.3 Correction Factors

The costs and benefits are not equally distributed through the lifetime of the abatement measures. To be able to compare the costs of each measure, annualized present values (annual costs) have been computed. In these calculations a discount rate of 7% per year and a lifetime of 30 years have been assumed. In Table 11.3 the costs and side effects are given in annualized present values.

All the abatement measures considered in this study will reach full effect by the year 2000, if they are implemented as planned in 1990. The attribute values x_1 - x_8 in Table 11.2 refer to the full effect in the year 2000 for all measures. Some of the countermeasures give the full effect immediately, others will have the full effect after about 10 years. For all abatement measures the full effect is assumed for the last 20 years of the economic lifetime of the measures. To account for the 10-year transition period when the full effect is not obtained for all measures, a correction factor is introduced. The correction factor for a measure is obtained by calculating the present value of its benefits using a discount rate of 7% per year, and then dividing by the present value under the assumption of full effect in all 30 years. When the correction factor is multiplied by the full effect estimate for the year 2000, one gets the average effect per year, or the "annualized" values, for these attributes over the 30-year lifetime. The values of the correction factors for the abatement measures are given in Table 11.2.

6 VALUATION OF THE IMPACT OF THE COUNTERMEASURES

To evaluate and rank the countermeasures, some criterion or an evaluation model is needed. For this purpose there exist a number of different models, including multiattribute utility theory (MAUT), cost/benefit analysis, cost-effectiveness analysis, and contingent cost-benefit analysis (see Trønnes et al., 1986). These models vary in complexity and need for subjective value judgments. The SFT was strongly in favour of a relatively simple evaluation model that would be intuitively easy to understand. The reasons for this point of view were that the model should be easy to communicate to the decision-makers, be easily accepted by the decision-makers and that it should not be too demanding for the decision-makers to make the necessary value judgments and trade-offs between the objectives needed to arrive at a conclusion.

6.1 The Evaluation Model

A commonly used variant of the evaluation model in multiattribute utility analysis is the additive form of the utility function U (Merkhofer and Keeney, 1987; Anadalingam, 1987):

$$U(x) = \sum_{i=1}^{n} k_i u_i(x_i), \tag{11.4}$$

where the k_i's are scaling factors representing value trade-offs between units of the corresponding attributes and the u_i's are the utility functions for the attributes. The use of equation (11.4) requires participation from the decision-makers in determining both the utility functions and the scaling factors.

The evaluation model used in this analysis, shown in equation (11.5), differs from equation (11.4) in two aspects: the attribute values are used directly instead of determining the utility functions for each attribute, (this simplification means that we assume a linear relationship between the attribute values and the utility), and the benefit/cost ratios have been used to evaluate the abatement measures instead of the utility. The evaluation model then becomes:

$$R = B/C = \left(\sum_{i=1}^{9} k_i x_i \right) / \left(k_{10} x_{10} + k_{11} x_{11} \right) \tag{11.5}$$

where R is the benefit/cost ratio, and B and C are the benefits and costs respectively, measured in some common unit. The k_i's (i = 1, 2, ... 11) are weights representing a valuation of the attributes into a common unit, and the x_i's are the attribute values. Alternatively the net benefit, N, given by:

$$N = B - C, \tag{11.6}$$

can be used to evaluate the abatement measures.

The motivation for using the attribute values directly is to simplify the model and to avoid the need for asking the decision-makers to determine the utility functions for each attribute. The abatement measures considered here do not represent the final decision alternatives. The measures differ very much in size (costs and benefits), and a small measure will then have a small utility even if its benefit per unit cost is very large. Since the final decision alternatives here will be a combination of abatement measures, the first objective is to identify the measure's effectiveness in terms of ability to reduce pollution effects per unit cost. Therefore, the benefit/cost ratio has been used to evaluate the abatement measures instead of the net benefit. The net benefits for the measures, however, have also been calculated.

The benefit/cost ratio, R, is used to rank the abatement measures. Further, when R>1 for an abatement measure, the benefit is greater than the cost indicating that it may be useful to implement the measure.

6.2   Valuations and Trade-offs

Before the evaluation model can be used, the weights, k_i, must to be determined. This is the subjective part of the analysis, and one would expect different interest groups to come up with different sets of weights, reflecting their different preferences. It is therefore necessary for the decision-makers to participate in the valuation process.

Even if this part of the analysis is subjective, one will find some guidance in the literature and in earlier practice with similar decisions as to what can be considered reasonable values of the weights. The SFT worked out a set of weights based on their preferences and references to earlier works (Syversen, 1987). This list of weights is given in Table 11.4.

The attribute weights in Table 11.4 have been normalized so that 1000 NOK/year have been given the weight of 1. Among the exposure attributes the weights for NO_x and CO exposure are two and three times as large as for SO_2 exposure, respectively. Syversen (1987) argues that the reason for this is that a greater portion of those exposed to NO_x and CO in Oslo are exposed to concentrations much higher than the guidelines than is the case for SO_2. In addition, exposure to concentrations just above the guidelines for SO_2 and NO_x will affect only the most sensitive groups of the population, while for exposure to CO all persons will be affected. Further, it is argued that exposure to suspended particulate matter (SPM) should be given a high weight relative to SO_2 because it is assumed that SPM contains hydrocarbons and other organic compounds that contribute to elevated risks for lung cancer.

The weight corresponding to attribute x_1 is 1.89. This means that it is considered to be 1.89 times more important to obtain a reduction of one in the number of persons exposed to SO_2 (for one year), than it is to save one thousand NOK. In other words, one is willing to pay 1890 NOK to have one person less exposed to SO_2 for one year. The other attribute weights in Table 11.4 can be given similar interpretations.

Using Tables 11.2, 11.3, and 11.4 it is possible to calculate the total damage caused by the air pollution in Oslo. Without implementing any of the abatement measures considered here, the total damages, TD, in the year 2000 will be:

$$TD = (\sum_{i=1}^{8} k_i x_i)$$

$$= (1.89 * 185\,000 + 4.15 * 150\,000 + 3.77 * 12\,000 + 5.66 * 0$$

$$+ 0.23 * 230\,000 + 0.23 * 306\,000 + 2.45 * 5\,160 + 0.25 * 7\,800) * 10^3 \text{ NOK}$$

$$= 1\,153\,372 * 10^3 \text{ NOK} \tag{11.7}$$

In a similar manner the benefit/cost ratios of each abatement measure can be calculated using equation (11.5). Results from these calculations are shown in the next section.

6.3   Elicitation of Decision-Makers´ Preferences

A group of decision-makers (local politicians) were invited to a session to elicit their preferences in terms of attribute weights. Initially the decision-makers were reluctant to make the necessary value trade-offs. The set of weights given in Table 11.4 was, therefore, presented and explained by representatives of the SFT. These values served as starting values for the decision-makers to give their own preferences. The new sets of weights were put into a computer model that calculated new benefit /cost ratios and presented new rankinglists of the abatement measures. The session more or less developed into a sensitivity analysis of the weights. Results of this sensitivity analysis are discussed in section 9.

Table 11.4 SFT´s Attribute Weights per Unit of the Attributes

General concern	Attribute		Unit[1]	Attribute weight[2]
Exposure	SO_2	(x_1)	persons exposed to SO_2	1.89
	SPM	(x_2)	persons exposed to SPM	4.15
	NO_x	(x_3)	persons exposed to NO_x	3.77
	CO	(x_4)	persons exposed to CO	5.66
Well-being	Dust	(x_5)	persons afflicted by dust	0.23
	Odour	(x_6)	persons afflicted by odour	0.23
Acidification	SO_2	(x_7)	tons of SO_2 emitted / year	2.45
	NO_x	(x_8)	tons of NO_x emitted / year	0.25
Side effects		(x_9)	10^3 NOK / year	1.00
Costs	Private	(x_{10})	"	1.00
	Public	(x_{11})	"	1.00

[1] "Persons exposed" in this table refers to exposure to concentrations above the guidelines.
[2] Attribute weights have been normalized to give 10^3 NOK/year the weight of 1.

7 RESULTS

The benefit/cost ratio for each countermeasure was calculated according to equation (11.5). In Table 11.5 below the benefit/cost ratios (and the net benefits) for the counter-measures are given based on the SFT´s preferences, given in Table 11.4.

When an abatement measure is implemented, it will, in general, affect the benefit/cost ratios of the other measures. For example, the implementation of some measures may reduce the concentrations in one part of the city to levels below the guidelines. Implementation of another measure, then, will not contribute to a further improvement of the situation, and the benefit/cost ratio will be reduced accordingly. Table 11.5 provides the benefit/cost ratios and net benefits for each measure when no other measures have been implemented.

8 DEALING WITH UNCERTAINTIES

Section 5 described how the effects of the abatement measures were estimated. As an example, one may recall that the exposure assessment, i.e., assessment of the number of persons that are exposed to concentrations above the guidelines, comprises the following steps: assessment of the reduction in the emissions, dispersion modeling, and assessment of the geographical distribution of the urban population. Uncertainties are introduced at all these steps. Stochastic simulation or Monte Carlo simulation can be used to estimate the uncertainties in the number of persons exposed when uncertainties in the input variables are given in terms of probability distributions. (Chapter 2 of this book provides an illustration of the Monte Carlo simulation method.) Similarly, uncertainties will be introduced in the quantification of the attributes for well-being, acidification, side effects, and costs. The uncertainties will vary among the abatement measures. Some of the measures consist of carrying out information campaigns. For such measures significant additional uncertainty in terms of public responses to the campaigns will be introduced.

Uncertainties in the effect attributes of the abatement measures will result in uncertainties in the benefit/cost ratios given in Table 11.5. These uncertainties have to be quantified when considering the questions of whether the benefits are significantly greater than the costs of an abatement measure, or whether one measure is significantly better than another. In this section a method for estimating the uncertainty in the benefit/cost ratio will be described.

Table 11.5 Ranking of the Abatement Measures According to the Benefit/Cost Ratio

Rank no.	Abatement measure No.	Description	Benefit/Cost ratio	(Net Benefit)
1	5.2	Buses, better maintenance to reduce exhaust emissions	52.6	(21.6)
2	3.2	Diesel vehicles, better maintenance	29.2	(141.0)
3	10.7	Usage of low-sulphur oil under inversions	16.4	(109.3)
4	3.4	Diesel cars, reduced exhaust emissions	15.7	(41.2)
5	5.1	Buses, "braking energy"	8.4	(7.4)
6	3.6	Diesel trucks, reduced exhaust emissions	7.2	(186.0)
7	3.3	Prohibition on idle running of cars	6.4	(0.1)
8	5.5	Buses, new diesel engines	6.3	(10.6)
9	5.3	Buses, installation of equipment to reduce exhaust emissions	5.9	(4.3)
10	9.1	Reduced usage of wood for space heating	5.5	(103.5)
11	6.1	Electricity substitutes for fossil fuel	5.3	(15.1)
12	10.1a	Reduced sulphur content in low sulphur oil	4.7	(373.7)
13	9.4	Prohibition on burning of garden debris	4.6	(4.9)
14	9.2	Oil-based heating installations, better maintenance	3.9	(58.0)
15	10.1b	Reduced sulphur content in oil used for space heating	3.5	(50.0)
16	10.3	Max. 0.5% sulphur in certain oiltypes	3.4	(55.2)
17	1&2	Reduced private car usage	3.3	(577.3)
18	3.5	Heavy diesel vehicles, reduced exhaust emissions	2.8	(235.8)
19	4.2	Cleaning of roads	2.7	(3.4)
20	8.1	Energy saving measures	2.6	(6.9)
21	10.6	Usage of electricity under inversions	2.4	(57.4)
22	4.1b	Taxes on usage of studded tires	2.1	(176.0)
23	5.6	Buses, usage of gas fuel	1.9	(10.8)
24	10.4	Prohibition on oil with high sulphur content	1.6	(21.0)
25	5.4	Buses, usage of methanol fuel	1.4	(8.0)
26	7D	Remote heating, oil-based	1.4	(123.2)
27	4.1a	Prohibition on usage of studded tires	1.4	(213.6)
28	7F	Remote heating, based on oil and heat pump	1.4	(124.8)
29	9.3	Reduced emissions from industry	1.2	(1.5)
30	6.2	Taxes on oil, increased electricity-based heating	1.2	(0.6)
31	7I	Remote heating, based on oil, gas and heat pump	1.0	(0.0)
32	11.0	Vegetation screens	0.4	(-12.2)

8.1 Assessing Uncertainties

In this analysis no detailed quantification of the uncertainties in the effects estimates of the abatement measures has been carried out. At the start of the project, however, there were plans to do so. There are several reasons why the uncertainty analysis has been omitted: no quantitative data are available that can be used to estimate uncertainties statistically; many people involved in the quantification have little experience with probability theory; quantifying uncertainties means more numbers to deal with, etc. It is a paradox that when "experts" feel there is great uncertainty about a quantity, one often ends up with just a point estimate because they are reluctant to quantify the uncertainties.

However, it is important to get an idea of the magnitude of the uncertainties in the benefit/cost ratio and what factors contribute to the uncertainties. In many cases the only method to quantify the uncertainties is by subjective expert assessments.

To get an impression of the uncertainties in the benefit/cost ratios of the abatement measures considered, measure 9.2, "better maintenance of oil-based heating installations," was chosen for further investigation. The uncertainties in the effects of this abatement measure were assessed subjectively in terms of fractiles in their probability distributions. In the assessment the 11 attributes were grouped into 5 different categories, one for each of the general objectives: exposure, well-being, acidification, side effects, and costs. The relative uncertainty was assumed to be the same in each category. Table 11.6 shows judgmentally estimated relative fractiles in the probability distribution for these five categories.

Table 11.6 Relative Fractiles of the Probability Distributions, Given in Percentages of the Median Value, for the Five Uncertainty Categories

Uncertainty category	Attri- butes	Fractile						
		0%	5%	20%	50%	80%	95%	100%
Exposure	x_1 - x_4	25	50	70	100	130	150	195
Well-being	x_5 - x_6	40	60	80	100	125	145	180
Acidification	x_7 - x_8	60	80	90	100	110	120	140
Side effects	x_9	50	70	85	100	115	130	150
Costs	x_{10} - x_{11}	70	85	90	100	120	140	170

In Table 11.6 the fractiles are given as percentages of the median values. To find the actual fractiles for a given attribute, one has to multiply its median value given in Table 11.4 by the relative fractile value from Table 11.6. As an exampl, the 95% fractile of attribute x_1 of abatement measure 9.2 becomes 16,100 persons exposed * 150% = 24,150 persons exposed.

From Table 11.6 one can see that the exposure data is the category with the highest degree of uncertainty. This is reasonable because uncertainties are introduced at many steps in the estimation of these data.

The cumulative probability distributions for the uncertainty categories are plotted in Figure 11.2. Figure 11.2 is based on the fractile values given in Table 11.6; fractile values between those given in the table are obtained by linear interpolation.

8.2   Monte Carlo Simulation

The uncertainties in the benefit/cost ratio resulting from the uncertainty distributions given previously, can be estimated using stochastic simulations (or Monte Carlo simulations). The simulation procedure consists of two steps: attribute values are drawn stochastically according to their probability distributions, and the benefit/cost ratio is then calculated from equation (11.5). These two steps are replicated many times each resulting in a new "observation" of the benefit/cost ratio. Based on these observations, the mean, standard deviation, confidence intervals, etc., can be calculated, and the probability distribution can be plotted. The Monte Carlo simulation method is further explained in Chapter 2 of this book. A computer program was developed to perform the simulations.

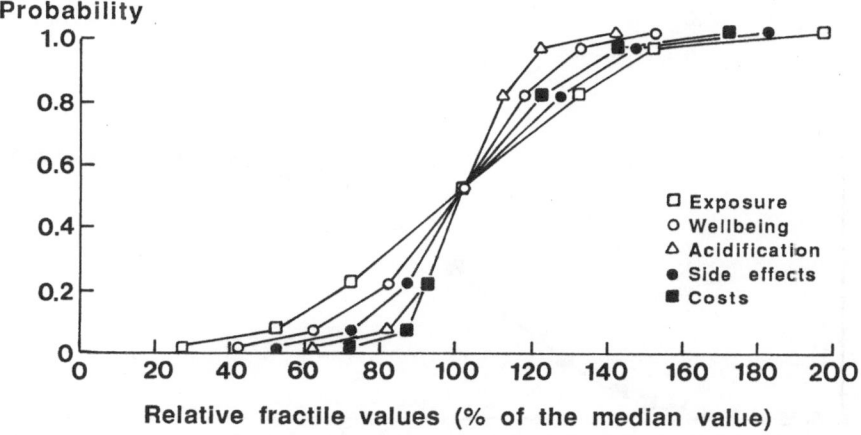

Figure 11.2   Cumulative frequency distributions for the five uncertainty
categories

In the present simulation 5000 sets of attribute values were used. For each replication of the simulation model, the "observations" of the benefit/cost ratio were sampled into intervals of length 0.2. Using the results from this sampling, one can plot the frequency distribution and cumulative frequency distribution of the benefit/cost ratio. These two plots are shown in Figure 11.3 and 11.4, respectively.

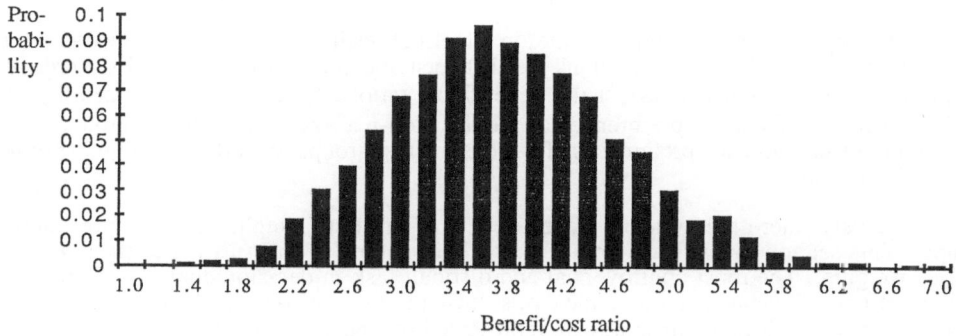

Figure 11.3   Frequency distribution for the benefit/cost ratio of abatement
measure no. 9.2

The simulated mean value of the benefit/cost ratio for abatement measure no. 9.2 is 3.78, the simulated standard deviation is 0.86.

From the sampling results (or from Figure 11.4), one can see that the probability is about 0.1 that the benefit/cost ratio is less than 2.7, and about 0.9 that the benefit/cost ratio is less than 4.9. An 80% confidence interval for the benefit/cost ratio of abatement measure 9.2 is, therefore, approximately [2.7 - 4.9]. In Table 11.5 one can see that altogether 8 of the abatement measures have benefit/cost ratios within this interval. This indicates that the uncertainty in the benefit/cost ratio of measure 9.2 is so large that one cannot conclude that measure 9.2 is significantly better than the measures with a slightly lower rank in Table 11.5, and vice versa. Further, the results from the simulations can be used to find the probability that the measures are not worth implementing, that is, the probability that the benefit/cost ratio is less than 1. Other confidence intervals and fractile values can also be found or calculated from Figure 11.4.

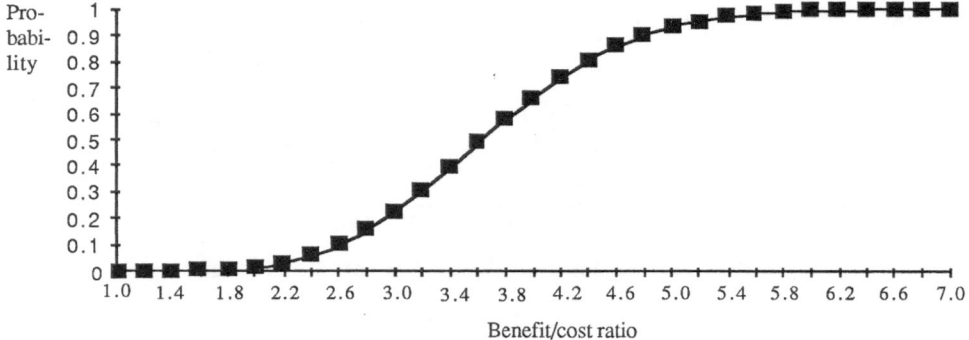

Figure 11.4   Cumulative frequency distribution for the benefit/cost ratio of abatement measure no. 9.2

9   SENSITIVITY ANALYSIS

Section 8 explained how uncertainties in the effects estimates affected the uncertainty in the benefit/cost ratio which is used to rank the abatement measures. The set of attribute weights, k_i, that one is using will also affect the benefit/cost ratios. Section 6 emphasized that the attribute weights reflect the preferences of the decision-makers or the interest groups. It is interesting to study how the preferences of various interest groups will affect the ranking of the abatement measures.

Obviously, there are as many different sets of attribute weights as there are interest groups. This section investigates how sensitive the ranking of measures is to the choice of various weights. The most controversial and difficult task in decision analysis of pollution problems often is to valuate, in economic terms, attributes that are concerned with human health. In the present analysis these attributes are represented by the exposure attributes x_1 - x_4. Two additional sets of attribute weights were constructed. In the first set the weights k_1 - k_4 were doubled compared to the weights given in Table 11.4, and in the second set the weights k_1 - k_4

Table 11.7 Ranking of the Abatement Measures for Additional Sets of Attribute Weights

| Abatement measure | | Rank no. | | | |
No.	Description	Original set	Set 1	Set 2	Set 3
5.2	Buses, better maintenance to reduce exhaust emissions	(1)	(1)	(1)	(1)
3.2	Diesel vehicles, better maintenance	(2)	(2)	(2)	(2)
10.7	Usage of low-sulphur oil under inversions	(3)	(3)	(4)	(3)
3.4	Diesel cars, reduced exhaust emissions	(4)	(4)	(3)	(4)
5.1	Buses, "braking energy"	(5)	(5)	(6)	(5)
3.6	Diesel trucks, reduced exhaust emissions	(6)	(6)	(7)	(6)
3.3	Prohibition on idle running of cars	(7)	(13)	(5)	(7)
5.5	Buses, new diesel engines	(8)	(7)	(10)	(9)
5.3	Buses, installation of equipment to reduce exhaust emissions	(9)	(8)	(12)	(11)
9.1	Reduced usage of wood for space heating	(10)	(9)	(15)	(13)
6.1	Electricity substitutes for fossil fuel	(11)	(10)	(14)	(12)
10.1a	Reduced sulphur content in low sulphur oil	(12)	(11)	(13)	(14)
9.4	Prohibition on burning of garden debris	(13)	(14)	(8)	(15)
9.2	Oil-based heating installations, better maintenance	(14)	(16)	(11)	(10)
10.1b	Reduced sulphur content in oil used for space heating	(15)	(15)	(17)	(17)
10.3	Max. 0.5% sulphur in certain oiltypes	(16)	(12)	(19)	(18)
1&2	Reduced private car usage	(17)	(20)	(9)	(8)
3.5	Heavy diesel vehicles, reduced exhaust emissions	(18)	(17)	(21)	(20)
4.2	Cleaning of roads	(19)	(24)	(16)	(21)
8.1	Energy saving measures	(20)	(21)	(18)	(16)
10.6	Usage of electricity under inversions	(21)	(18)	(24)	(22)
4.1b	Taxes on usage of studded tires	(22)	(28)	(20)	(19)
5.6	Buses, usage of gas fuel	(23)	(19)	(27)*	(26)*
10.4	Prohibition on oil with high sulphur content	(24)	(22)	(29)*	(28)*
5.4	Buses, usage of methanol fuel	(25)	(23)	(30)*	(29)*
7D	Remote heating, oil-based	(26)	(26)	(25)	(23)
4.1a	Prohibition on usage of studded tires	(27)	(31)	(23)	(24)
7F	Remote heating, based on oil and heat pump	(28)	(27)	(26)	(25)
9.3	Reduced emissions from industry	(29)	(30)	(22)	(30)*
6.2	Taxes on oil, increased electricity-based heating	(30)	(25)	(31)*	(31)*
7I	Remote heating, based on oil, gas and heat pump	(31)	(29)	(28)*	(27)*
11.0	Vegetation screens	(32)*	(32)*	(32)*	(32)*

* Indicates that for this abatement measure the benefits are less than the costs

were multiplied by 0.5. For both sets the benefit/cost ratios were calculated and the abatement measures ranked. The results of these rankings are shown in Table 11.7.

Economy will always be important when the decision-makers are to make the final judgments about what abatement measures should be implemented. It is, therefore, interesting to study how the ranking will change when the weights of the attributes measured in monetary terms, x_9 - x_{11}, are increased. In a third set the weights k_9 - k_{11} have been multiplied by 2 relative to the weights given in Table 11.4. The results of the ranking are shown in Table 11.7.

The general impression from Table 11.7 is that the rankings are not altered dramatically; however, there are some exceptions. If we compare sets 1 and 2 in the table, the biggest differences occur for measures where the benefits mainly occur either in the exposure attributes or mainly in the other attributes. One can see that measure 3.3 improves its ranking from 13 to 5 when the attribute set changes from 1 to 2. The reason is that 3.3 is a very small measure that costs only 0.022 million NOK. This measure is assumed to contribute to increased benefit only by improving the well-being attributes (cfr. Figure 11.1), having no effect on for example the exposure attributes, so when exposure reduces its importance in set 2 the relative importance of the other attributes increases and the measure gets a higher ranking. The opposite happens with measure 11.3.

If we look at set 3, where the monetary values have a higher weight, the deviations from the original set are less than for sets 1 and 2. The most significant exception here is measure 1&2 which increases its ranking from 17 with the original set to 8 with set 3. The reason is that for this measure the side effects given in monetary terms are greater than the costs, so increasing the weight put on monetary values increases the benefit more than the costs.

The absolute values of the benefit/cost ratios will, of course, be very sensitive to the choice of alternative sets of weights. This will influence the number of measures with benefit/cost ratios of less than 1. Such measures are indicated by an asterisk in Table 11.7.

10   CLOSING REMARKS

The ultimate goal of this analysis has been to estimate the costs and benefits of a number of feasible measures to reduce air pollution in Oslo. Despite limited time and resources available for the project, all of the necessary steps in the analysis were carried out and useful results were obtained about the countermeasures. The project resulted in a ranking of the abatement measures according to their benefit/cost ratio (see Table 11.5). This ranking, together with information about the benefits and costs of selected combinations of the abatement measures, was presented to the decision-makers and politicians in Oslo and to the Norwegian government in September 1987.

On several occasions during the project, decisions had to be made regarding the level of detail in the data collection and what models and methods should be used. Limited resources resulted in some necessary shortcuts; thus, the final analysis was methodologically less comprehensive than planned at the start of the project. Some brief comments follow on two of the shortcuts. First, one of the main motivations for performing this analysis was the belief that the air pollution in Oslo could cause adverse health effects in humans (lung cancer, obstructive lung diseases, angina attack, etc.). However, lack of good dose-effect and dose-response models made the quantification of these effects difficult and controversial. It was therefore decided to use the exposure assessments as the final attributes for describing the health situation. One consequence is that the decision-makers are left with the problem of making trade-offs between objectives like reducing exposure for SO_2 on one hand and for NO_x on the other, etc., and between reducing exposure versus other attributes. The question that naturally is raised by the decision-makers will be which of the pollution components are most important for the development of a given health effect, and how many incidences can be expected at a given concentration level. When decision-makers were asked about alternative sets of attribute weights, they naturally had problems in coming up with good suggestions. This may indicate that it would be better to use existing medical expertise as far as possible, even if it is uncertain in this field, than to leave the question totally to the decision-makers.

Second, the abatement measures considered in this analysis vary considerably in size, and some are restricted to just a few pollution components and geographical areas. It is therefore important to identify optimal combinations of abatement measures and consider these combinations of measures as the final decision alternatives. (This problem will occur in many pollution analyses where the approach is to implement measures that will improve the situation in the affected area.) However, to be able to identify optimal combinations and calculate their benefit/cost ratios, one has to describe certain dependencies between the measures. In this project the available resources did not permit establishing a model for these dependencies, and the combination of measures had to be done on a rather ad hoc basis.

11   ACKNOWLEDGEMENT

The abatement measures considered in this paper were defined by the Norwegian State Pollution Control Authority (SFT) in cooperation with other institutions and organizations working in the relevant fields. The effects of the abatement measures have been estimated also by the SFT in cooperation with related institutions. The Norwegian Institute for Air Research (NILU) has contributed a lot to the exposure assessments by performing calculations on their air dispersion models. This paper was written before the project was completed. Some of the results presented here may, therefore, be changed. We are grateful to the SFT for putting their material at our disposal and for many stimulating and helpful discussions with Terje Kronen and Trond Syvertsen of the SFT. Hans M. Seip and Anders B. Heiberg of the Center for Industrial Research have given many valuable comments on several versions of this paper.

12   REFERENCES

Anadalingam, G., 1987, A multiple criteria decision analytic approach for evaluating acid rain policy choices, European Journal of Operations Research, pp. 336-352.

Aune, T.,1982, "Health Effects from Air Pollution in Oslo," (in Norwegian), SFT-report no. 41, The Norwegian State Pollution Control Authority, Oslo.

Keeney, R.L., and Raiffa, H.F., 1976, "Decisions with Multiple Objectives: Preferences and Value Trade-offs," John Wiley & Sons, New York.

Merkhofer, M.W., and Keeney, R.L., 1987, A multiattribute utility analysis of alternative sites for the disposal of nuclear waste, Risk Analysis, 7(2) : 173-194.

Norsk Opinionsinstitutt, 1986, "Inquiry about Air Pollution Nuisance in Oslo," Norsk Opinionsinstitutt, Oslo.

Overrein, L.N., Seip, H.M., and Tollan, A., 1980, "Acid Precipitation - Effects on Forest and Fish," Final report of the SNSF-project 1972-1980, NIVA, Oslo.

Grønskei, K.E., Gram, F., and Larssen, S., 1982, "Calculation of Dispersion and Exposure for Some Air Pollution Components in Oslo," (in Norwegian), NILU/OR 8/82, The Norwegian Institute for Air Research, Lillestrøm.

SFT, 1985, "Further Reduction of Air Pollution in Oslo," (in Norwegian), The Norwegian State Pollution Control Authority, Oslo.

SFT, 1982, "Air Pollution: Health and Environmental Effects," (in Norwegian), SFT-report no. 38, The Norwegian State Pollution Control Authority, Oslo.

Syversen, T., 1987, "Comparing the Utility of Various Impacts of Measures to Reduce Air Pollution," (in Norwegian), The Norwegian State Pollution Control Authority, Oslo.

Trønnes, D.H., Heiberg, A.B., and Seip, H.M., 1986, Decision making in pollution control, in: "Risk and Reason: Risk Assessment in Relation to Environmental Mutagens and Carcinogens," pp. 127-140, Alan R. Liss, Inc., New York.

SUMMARY, CONCLUSIONS AND RECOMMENDATIONS

A. Heiberg and H.M. Seip

Center for Industrial Research
Oslo, Norway

1   INTRODUCTION

It is not possible to cover all aspects of risk assessment and risk
management in a study of this kind. Nevertheless, many important topics
were covered during the study. In general, the participants contributed
projects in which they were involved when the Pilot Study was initiated.
As a result, each sub-project was carried out primarily as a national
undertaking. However, the Pilot Study meetings that were arranged and
visits between Study Group members had in some cases considerable
influence on the project development.

Although the Pilot Study was initially aimed at risk management, it
soon became apparent that it was necessary to include aspects of risk
assessment (hazard identification, dose-response assessment, exposure
assessment, risk characterization) as well. It should be noted that while
risk assessment in principle can be carried out objectively, risk
management, which includes the process of weighing policy alternatives,
necessarily must involve preferences and attitudes.

The previous chapters of this book describe various aspects of risk
assessment and risk management of chemicals in the environment. Brief
summaries of these chapters are given in the next section. General
conclusions based on the material presented are formulated in section 3,
which also includes the Study Group's recommendations to the CCMS.

2   SUMMARY OF THE INDIVIDUAL CHAPTERS IN THE REPORT

The first chapter outlines the background for the project. It is
pointed out that although basic concepts of risk assessment and risk
management have been applied for a long time, the need for detailed
technical and scientific methodologies has only recently been realized.
The chapter presents very briefly important components of risk assessment
and risk management related to control of chemicals. Various quantitative
decision analysis techniques, suitable for managing the effects of
chemical exposure, are briefly described, and the importance of
interaction between the decision-maker and the groups affected by the
problem of concern is underlined.

Chapter 2 discusses methods and uncertainties in the quantification of health risks due to chemicals with emphasis on the problem of establishing dose-response relationships. The chapter reviews the most widely used methods for quantifying the relationship between health risk and chemical exposure, including epidemilogical studies and short- and long-term laboratory tests. Special attention is given to the limitations and uncertainties associated with the various methods. The use of stochastic simulations in health risk quantification is illustrated by estimating the number of excess angina attacks in an urban population caused by exposure to carbon monoxide. The result is a frequency distribution of the number of angina attacks, which, in addition to giving the most likely number of attacks, also illustrates the uncertainty in the number obtained.

Chapter 3 first gives a brief survey of the various elements of an environmental risk analysis, i.e., release, transport and fate of chemicals in the environment; exposure routes; and dose-response relationships. The discussion is mainly based on experience gained in the Netherlands. It is pointed out that a detailed environmental risk analysis is often not possible and indeed not required under all circumstances. In order to support decision making, a simplified approach may be sufficient. This point is illustrated by a study of risks of soil contamination by leakage from underground storage tanks and a discussion of risks of soil contamination from industrial activities.

In recent years, the United States has moved toward a risk assessment/risk reduction framework to make regulatory decisions, in particular regarding human health. The United States Environmental Protection Agency (EPA), in order to ensure the quality and consistency of the risk assessment component of those decisions, has developed risk assessment guidelines for carcinogenicity, mutagenicity, developmental toxicity, toxicity of chemical mixtures, and exposure. The main points of these guidelines are presented in Chapter 4, along with other guidelines developed in the United States for the assessment and management of health risks due to chemicals. The risk assessment of dichloromethane is discussed as an example of a recent effort in this area. The conclusion of the study is that the compound should be classified as a probable human carcinogen.

The discussion of health risks in Chapter 4 is followed by a similar discussion of EPA's treatment of ecological risks (Chapter 5). There are many problems associated with assessments of ecological risk. One is the paucity of models and predictive methods for the very complex situations one often has to face. Because of the lack of empirical data, the validation of methods used to predict ecosystem response is difficult. In addition, it is not fully understood how to use such data to conduct ecological risk assessment. Efforts recently have begun in the United States at the EPA to generate data, develop and validate models, and develop ecological risk assessment guidelines.

Although cost-benefit analysis and related methods can be a valuable tool for incorporating scientific evidence and economic information into the legal and administrative process, its use as a basis for formulating and implementing environmental policy to manage the adverse effects of chemical exposure, remains controversial. Chapter 6 describes a study of the attitudes within the EPA toward using risk assessment and economic analysis in formulating environmental policies. Data for this study were obtained through a series of interviews conducted with 34 senior officials of the EPA, whose responsibilities involved assessing risks or developing policies to prevent or mitigate environmental damages. A large majority of those interviewed were in favour of applying risk assessment and cost-benefit analysis (76.5 and 64.7%, respectively), but most of them (70.0%) disagreed with the idea that one should place an economic value on human life. Educational background rather than office or job responsibilities appears to play the dominant role in shaping attitudes towards the utility of risk analysis, cost-benefit analysis, methods for valuation of human life, and distribution of risk.

The estimation of exposure of humans, other animals or plants to pollutants requires quantitative evaluation of the processes in which the pollutants take part in the environment. In most cases such evaluations will have to be based on mathematical models describing the fate and transport of the chemical investigated. Chapter 7 describes an empirical model for the evaluation of the mobility and degradation of pesticides in soils. The model takes into account the following processes: adsorption and partitioning between phases, capillary and gravitational water flow, plant uptake, and degradation of the compound. The model was validated using results from a field experiment with atrazine combined with the relevant physical-chemical data. Except for the initial transitory phase following the addition of the compound to the soil surface, where the model overestimated the pesticide concentration in the upper soil layer, the agreement between observed and predicted concentrations was fairly good.

The fate of chemicals is also dealt with in Chapter 8. The experimental data available on the environmental contamination that resulted from the Seveso accident in Italy, are examined in order to determine the environmental behaviour of 2,3,7,8-tetrachlorodibenzo-p-dioxin (TCDD) and of other organic chemicals with similar characteristics. The study shows that the TCDD levels detected in the environment may be regarded as the consequence of a particular release pattern and an environmental mobility that is mainly due to an initial mechanical transport by water. The mobility is progressively reduced by TCDD's binding to organic carbon in the soil. Even though the solubility of TCDD in water is extremely low, dissolved TCDD is one cause of plant contamination in the affected area; another cause is deposition of contaminated dust on plant tissues.

Chapter 9 addresses the mathematical and biological uncertainties involved in assessing a permissible level of lead in blood in humans. The concentration of free erythroporphyrin (FEP) in blood is used to determine a "No Observed Adverse Effect Level" (NOAEL) for lead. The NOAEL, for a given population, is defined as the concentration of lead in blood (PbB) at which the FEP concentration starts to increase significantly above the background level. In a study of a population of children abnormally exposed to dietary lead, an increase in FEP was found for PbB values greater than 24.3 µg/dl, with 95% confidence limits of 21.9 and 26.6 µg/dl. This result was obtained by means of the "hockey stick" regression method. Using this method the mathematical uncertainty of the NOAEL is reasonably small. Some biological uncertainty remains, however. For example, it is known that insufficient iron absorption from the gut increases the lead uptake. This and other commonly neglected biases should be taken into account when using the FEP-index in decision making. It is claimed that a further decrease of alkyl lead in gasoline is not likely to cause a substantial reduction in the number of people in Europe having a PbB value above the established NOAEL. Therefore, in dealing with the health risk posed by environmental lead contamination, other sources of lead intake should be given more consideration.

Chapters 10 and 11 both describe case studies illustrating the use of quantitative decision techniques in environmental risk management.

In Chapter 10 two different methods are described for estimating the societal benefits of implementing pollution control in the Kristiansand fjord, a heavily polluted fjord in southern Norway. In the first method the willingness to pay for achieving a practically full cleanup of the fjord was estimated through interviews carried out locally and among the entire Norwegian population. In the second method the local residents were asked how they would evaluate various limited meaures directed towards different parts of the pollution problem. Using the Simplified Multi-attribute Rating Technique (SMART), a utility function for the Kristiansand fjord, focusing on water clarity, recreational fishing and the soft bottom fauna, was constructed from the answers obtained in the poll. By means of this function the net benefit for a number of pollution abatement measures was calculated. Both methods employed provide useful means of evaluating pollution control actions. SMART, however, can offer the more detailed picture of people's preferences regarding environmental values.

Chapter 11 describes a cost-benefit study of measures to mitigate the air pollution in Oslo, Norway. The problem considered is very complicated. A large number of pollutants may be relevant to the problem, including sulphur dioxide, nitrogen oxides, carbon monoxide, organic compounds and suspended particulate matter. A total of 32 measures, directed towards different types of pollutants and different sources, were investigated with an aim to determining their relative cost-effectiveness. To be able to carry through the study, rather drastic simplifications in the approach were necessary compared to a complete cost-benefit analysis. Thus, instead of explicitly considering health effects only reduced exposure was estimated. The major goal formulated was to reduce the number of people exposed to concentrations above established guideline values for the relevant air pollutants. Apart from reduced exposure, four general objectives were identified: increase well-being, reduce acidification, increase positive and reduce negative side-effects, and reduce costs. Weights reflecting the relative importance of these objectives were worked out by a project group at the Norwegian State Pollution Control Authority. Using these weights together with estimates of the effects of the abatement measures investigated, the benefit/cost ratio of each measure was calculated and a ranking of the measures set up. The list of measures arrived at will be an important part of the basis for formulating and implementing air pollution control in Oslo.

3   CONCLUSIONS AND RECOMMENDATIONS

3.1   General Conclusions

The Pilot Study resulted in useful cooperation among groups in the participating countries. Such international and interdisciplinary collaboration is extremely important for further progress in the field of risk assessment and risk management related to environmental pollution.

The study shows that formal methods (e.g., cost-benefit analysis) may be useful tools in managing the risk posed by chemicals. Formal methods help to structure the problem of concern and provide excellent opportunities for communication between decision-makers and groups affected by the problem. Also, they may result in a ranking of decision alternatives, thereby helping the decision-makers to set priorities and to choose among available alternatives. Formal methods now seem to be gaining increasing acceptance by regulatory agencies responsible for formulating environmental policies.

In recent years considerable experience has been obtained in assessing and managing health risks. However, in most cases there is a serious lack of data both on exposure to chemicals and on the relationship between dose and effects. This complicates the use of quantitative methods in risk assessment and risk management. These problems are still much more pronounced for the effects of chemicals on ecosystems, especially for long-term exposure to low doses.

In several countries attempts have been made to standardize approaches for assessing and managing risks to human health and ecosystems, for example, by developing guidelines. Such guidelines may be useful; however, at this stage of development flexibility in choice of approaches to risk assessment and risk management is essential. In particular, the experience of using formal decision-analytic methods in environmental risk management is still too limited to justify the adoption of one particular method. The Study Group, therefore, will not give specific recommendations regarding which methods to use and has made no attempt to develop a detailed set of guidelines or a protocol for managing the effects of chemical exposure. However, two points should be made: First, detailed risk analysis is not required under all circumstances. In order to assist the decision-maker simplified approaches may be sufficient. Second, it should be a general requirement that the uncertainties accumulating from all steps in the analysis appear clearly in the presentation of the results.

There are a large number of research needs related to the risk assessment and risk management of chemicals in the environment. In particular, we will mention the following areas:

- An effort is needed to increase our understanding of harmful effects caused by chemicals on ecosystems.

- There is a need for improved models for chemical fate in assessing risks both to health and the environment.

- Exposure assessments are often highly uncertain and better methods and more data are required.

- Better dose-response relationships are required in many cases, in particular for long-term effects.

- Valuation of effects will often include commodities for which there are no market prices. Continued research on how to model and elicit people's preferences is of great importance. Emphasis should be placed on establishing methods that are comprehensible and easy to use and at the same time represent people's preferences in an adequate way.

- New ways to include the affected groups in the decision process should be tried.

3.2 Recommendations to the CCMS

The Study Group in its Summary Report to CCMS (see Appendix G) made a number of recommendations to CCMS.

The committee was invited to:

- Work for increased efforts to protect humans and the ecosystem from harmful pollutants.

- Encourage membership countries to reduce emissions of harmful chemicals to the environment.

- Encourage international cooperation and interdisciplinary research in risk assessment and risk management. One way to achieve this is by continuing to provide CCMS fellowships and by giving collaboratory research grants in this field.

- Encourage national and international agencies to support work in this field and to use risk assessment and risk management methods in their work.

- Provide for means to follow up this Pilot Study with additional studies in related areas, e.g., studies of dose-response relationships for effects of chemicals on ecosystems or methods for eliciting people's preferences.

- Contribute to improved information about the effects of chemicals on human health and the environment to the general public.

APPENDIX A - G

APPENDIX A: Names and addresses of Study Group members
(national representatives and project co-workers)

Pilot Study Director:

Hans Martin SEIP

Center for Industrial Research
P.O.Box 124 Blindern
0314 Oslo 3
Norway

Cornelus L. van DEELEN

TNO
Division of Technology for Society
Department of Industrial Safety
P.O.Box 342
7300 AH Apeldoorn
The Netherlands

James W. FALCO

Office of Environmental Processes
and Effects Research
United States Environmental
Protection Agency
Washington, D.C. 20460
USA

Anders B. FEIBERG

Center for Industrial Research
P.O.Box 124 Blindern
0314 Oslo 3
Norway

Karl-Gerhard HEM

Center for Industrial Research
P.O.Box 124 Blindern
0314 Oslo 3
Norway

Carlo LUPI

Istituto Superiore di Sanità
Viale Regina Elena 299
00161 Roma
Italy

Richard V. MORASKI

Office of Health and
Environmental Assessment
United States Environmental
Protection Agency
Washington, D.C. 20460
USA

James L. REGENS

Institute of Natural Resources
The University of Georgia
Rm 13 Ecology Building
Athens, Georgia 30602
USA

Désiré RONDIA   Laboratoire de toxicologie
                                       de l'environnement
                                       Institut de Chimie-B6
                                       Sart-Tilman
                                       Université de Liège
                                       B-4000 Liège
                                       Belgium

Francis SARTOR   Laboratoire de toxicologie
                                       de l'environnement
                                       Institut de Chimie-B6
                                       Sart-Tilman
                                       Université de Liège
                                       B-4000 Liège
                                       Belgium

Dag Helge TRØNNES   Center for Industrial Research
                                       P.O.Box 124 Blindern
                                       0314 Oslo 3
                                       Norway

Giovanni A. ZAPPONI   Istituto Superiore di Sanità
                                       Viale Regina Elena 299
                                       00161 Roma
                                       Italy

Note: In addition to those listed above, a number of other people
      attended one or more of the Study Group meetings, making
      valuable contributions to the discussion. These included:

W. van Heugten   TNO, The Netherlands

A. Jensen   IMSOR, Danish Technical School,
                                       Denmark

A. Linos   School of Medicine,
                                       University of Athens,
                                       Greece

R. Stern   The Dansih Welding Institute,
                                       Denmark

M. Vaillant   France

APPENDIX B: Names and addresses of recipients of CCMS
Fellowships who contributed to the Pilot Study

1984 FELLOWSHIPS:

C.F. Cavelli
Via Maffeo Pantaleoni 23
00191 Roma
Italy

D. Rondia
Laboratoire de toxicologie de l'environnement
Institut de Chimie-B6
Sart-Tilman
Université de Liège
B-4000 Liège
Belgium

1985 FELLOWSHIPS:

W. van Heugten/C.L. van Deelen[*]
TNO
Department of Industrial Safety
P.O.Box 342
7300 AH Apeldoorn
The Netherlands

J.L. Regens
Institute of Natural Resources
The University of Georgia
Rm 13 Ecology Building
Athens, Georgia 30602
USA

1986 FELLOWSHIPS:

C. Lupi
Istituto Superiore di Sanità
Viale Regina Elena 299
00161 Roma
Italy

J.L. Regens (extension, see above)

---

[*] In 1986 Dr. van Deelen replaced Dr. van Heugten as recipient of the
CCMS fellowship. This change was formally approved by the CCMS in
September 1986.

APPENDIX C: Study Group programme

Meeting	Dates	Location	Hosts
Inaugural	April 11, 1984	Oslo, Norway	Royal Norwegian Council for Scientific and Industrial Research
Second	March 12, 1985	Oslo, Norway	Center for Industrial Research
Third	Nov. 12-13, 1985	Washington D.C., USA	U.S. Environmental Protection Agency
Fourth	April 27-28, 1986	Rome, Italy	Istituto Superiore di Sanità
Fifth	Nov. 21, 1986	Liège, Belgium	University of Liège, Institute of Chemistry
Final	Oct. 22-23, 1987	Oslo, Norway	Center for Industrial Research

In addition several study visits between participating countries were arranged during the Pilot Study period.

APPENDIX D: Work done by recipients of CCMS Fellowships

C.F. Cavelli   Fellowship used for making a comprehensive
                                  review and comparison of existing formal
                                  decision-analytical tools, including
                                  discussions of various aspects of environ-
                                  mental decision-making. Part of the work
                                  has been incorporated into Chapter 1 of the
                                  present report.

W. van Heugten/   Fellowship used for travel in Europe in
C.L. van Deelen   connection with Study Group meetings.
                                  Dr. van Deelen is the author of the paper
                                  on methods for assessing the risk of
                                  environmental contamination (see Chapter 3).

C. Lupi   Author of the paper on modelling of the
                                  behaviour of pollutants in soil for risk
                                  assessment purposes (see Chapter 7).
                                  Co-author of the report on the environ-
                                  mental and health impact of the Seveso
                                  accident (see Chapter 8). Fellowship
                                  also used for study visits to Norway and
                                  the United States.

J.L. Regens   Author of the report on attitudes toward
                                  risk-benefit analysis for managing effects
                                  of chemical exposure (see Chapter 6).
                                  Fellowship also used for travel in con-
                                  nection with Study Group meetings and for
                                  study visits to Norway and the United
                                  Kingdom.

D. Rondia   Leader of the project on health risk from
                                  lead exposure and co-author of the report
                                  on mathematical and biological uncertain-
                                  ties in the assessment of a permissible
                                  blood lead concentration (see Chapter 9).
                                  Fellowship also used for travel in con-
                                  nection with Study Group meetings.

SAMPLE SURVEY ABOUT THE KRISTIANSAND FJORD

QUESTIONNAIRE - NATIONAL STUDY 2 x 250 PERSONS

VARIANT I: KRONER 200

(VARIANT II = VARIANT I, EXCEPT THAT ALL THE AMOUNTS ARE HALVED)

QUESTION XX1

Which factors do you think are most important when evaluating the natural environment along the coast? Try to grade each point on a scale from 1 to 10 where 10 means very important and 1 means little importance.

ONE MARK FOR EACH LINE START FROM THE BOTTOM EACH SECOND TIME	Very important	Little importance
	10 1	
	Results, nationwide study	

Good water transparency	8.7
No taste and smell from the water.........................	9.5
No health risk by swimming	9.7
No health risk by eating the fish	9.7
A lot of fish	8.4
The bottom flora and fauna are normal	9.2
The kelp and seaweed seems fresh and normal	9.2
The beach is tidy	9.6

QUESTION XX2

SHOW MAP

The following question is about the fjord area outside Kristiansand (south). The inner part of this fjord is one of the most polluted areas in Norway.

Certain cleaning efforts are being considered to improve the situation in all the areas that were mentioned in the previous question. After such an effort the water will be as clean as that near other larger towns along the coast. The fish that are caught in the fjord can again be eaten without risks.

The authorities are now trying to estimate the benefits of a cleaner fjord. These benefits may be expressed by people's willingness to pay to get a cleaner fjord. The answers to the next questions will therefore have influence on whether such measures will be introduced or not.

It is not easy to find out how much the measures will cost, and how the payment should be distributed. However, in the end the costs have to be paid by society, i.e., by the polluting industry and/or the local and governmental authorities.

Suppose the costs to clean the Kristiansand fjord were divided among all taxpayers in the whole of Norway by an extra tax in 1986. If this extra tax was 200 Norwegian Kroner (NOK) (ca. 28 U.S. dollars) for an average taxpayer, would you be willing to support this proposal?

Results, nationwide study
(Responses in per cent)

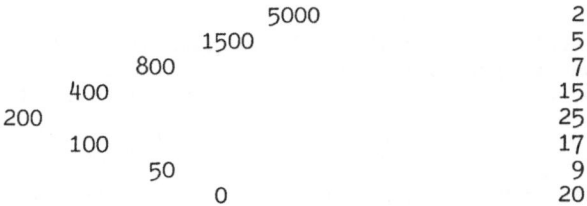

	Results
5000	2
1500	5
800	7
400	15
200	25
100	17
50	9
0	20

Instructions, questions XX2 and XX3:

Mark only <u>one</u> ring, around the <u>highest</u> amount the respondent (res) is willing to pay.

Example XX2:
a) If res is willing to pay 200 NOK, try 400 and say:
    "Suppose it will cost 400 NOK?"; and then if res is <u>not</u> willing to pay 400 NOK, put a ring at 200. If res <u>is</u> willing, try 800, and so on.

b) If res is <u>not</u> willing to pay 200 NOK, try 100 and say: "Suppose it will cost 100 NOK?" If yes, put a ring at 100, or else try 50, and so on.

QUESTION XX3

Another heavily polluted fjord is Frierfjorden, which is nearby Skien and Porsgrunn (two towns a little closer to Oslo).

Suppose that cleaning this fjord can be financed in exactly the same manner as the Kristiansand fjord, i.e., by an extra tax to pay for the costs. If the costs to clean <u>both</u> the Kristiansand fjord <u>and</u> Frierfjorden were divided among all taxpayers in Norway by an extra tax of 400 NOK, would you be willing to support this proposal?

	Results, nationwide study (Responses in per cent)
5000	2
1500	7
800	25
400	14
200	23
100	12
50	8
0	23

QUESTION XX4

Put only to res who have answered 50 NOK to question XX2 or XX3:
What is the reason for your low answer?

		Yes	No	% "yes" (nationwide study)
(1)	I think that industry should pay	☐	☐	57
(2)	Pollution is not that important to me	☐	☐	31
(3)	Other areas in the society are more important, e.g. schools, hospitals, etc	☐	☐	51
(4)	I think that I pay enough taxes already	☐	☐	51
(5)	Other reasons	☐	☐	34

QUESTION XX5

Put only to res who have answered 600 NOK or more to question XX2 or XX3:
What is the reason for your relatively high answer?

		Yes	No	% "yes" (nationwide study)
(1)	The fjord means a lot to me, I often use it	☐	☐	38
(2)	It is important to conserve the environment for society and for our descendants	☐	☐	92
(3)	Other reasons	☐	☐	38

QUESTION XX6

Do you have a house or a cottage in Kristiansand?   Yes   3
                                                      No   97

QUESTION XX7

Are you a member of any environmental   Yes   3
protection organization?   No   97

QUESTION XX8

Do you think the current total effort   Too small   65
to protect the environment is too small, Just right   26
just right, or should the effort to stop Should be reduced   1
pollution be reduced?

THE FOLLOWING QUESTIONS WERE ASKED IN THE LOCAL SURVEY ONLY:

To obtain satisfactory treatment and cleaning of the municipal sewage, it
is necessary to increase the public investments in this sector by about
100 per cent. One way to pay for these investments would be to increase
the local sewage excise tax in the municipalities in question. Note that
this replaces the payment arrangements in the previous questions. In this
case the sewage excise tax would increase to 500 NOK on a permanent basis
for an average family. Would you be willing to support this proposal?

Results
(Responses in per cent)

				Results (Responses in per cent)
			300	3
		1500		3
	1000			4
700				6
500				18
300				6
	200			8
		100		7
			0	45

(The other half of the sample started at 300 NOK)

QUESTION   XX9

Responses in per cent

Do you often use the Kristiansand fjord

	Yes, often	Yes, now and then	No, never
for swimming?	16	30	54
for boating or sailing?	18	29	53
for fishing?	9	15	74

QUESTION   XX10

Do you often eat fish that is caught in the Kristiansand fjord?

Responses in per cent

More than 10 times a year	☐	11
Between 5 and 10 times a year	☐	8
Less than 5 times a year	☐	12
No, never	☐	66

BACKGROUND VARIABLES (XX11) (SELECTED IN BOTH SURVEYS)

- Sex
- Age
- Family income
- Education
- Occupation
- Startpoint for the bids (1 or 2)

CONFIDENTIAL

INQUIRY ABOUT THE KRISTIANSAND FJORD 1987

As you may know, parts of the Kristiansand fjord are among the most polluted fjord areas in Norway. The authorities are interested in finding out what benefits a cleaner fjord would bring to the people in the area, and in which specific areas it is most important to make improvements. The answers to the following questions will therefore have some impact on evaluating which measures ought to be carried out.

EFFECTS OF CLEANUP ACTIONS

In this survey we will look at three different effects that could result from cleanup measures.

A Clearer water in Vesterhavn and Fiskåbukta

The amount of undissolved particles and algae determine the clarity of the water. Water clarity is quantified in terms of the secchi-disk depth; a larger secchi-disk depth implies clearer water.

Today the secchi-disk depth in Vesterhavn and Fiskåbukta varies considerably from place to place and time to time. At one place in Vesterhavn the secchi-disk depth has been observed to vary from 2.5 to 17.0 metres. The average depth in the summer was 7.1 metres. The Directorate of Public Health requires a secchi-disk depth of at least 2 to 3 metres for public bathing areas. Near Falconbridge Nikkelverk and at other places close to the city the water quality is significantly poorer than this.

An overall improvement of the secchi-disk depth in Vesterhavn-Fiskåbukta will give the area a more attractive appearance and make it more suitable for recreational purposes such as swimming, wind surfing, and other activities.

One way of achieving this is to stop the dumping of sewage into Vesterhavn by building a joint system for the western part of the city which will run out at Vestergapet.

B Improved recreational fishing due to reduction of pollution levels in fish and shellfish

Surveys carried out in 1982-85 revealed that fish in the Kristiansand fjord contain abnormally high concentrations of certain organic micropollutants. Some of these can pose a health risk. The highest concentrations have been found in fish from Vesterhavn. Fish containing larger amounts of organic micropollutants than normal have also been found further out in the fjord, including at Vestergapet. The Public Health Authorities in Kristiansand have considered the problem and have advised the public via the media not to eat fish from Vesterhavn. As for fish caught in the eastern and outer fjord areas, this warning applies especially to the consumption of fish liver. The whole of the flounder, in particular, is not thought to be fit for human consumption according to the health authorities.

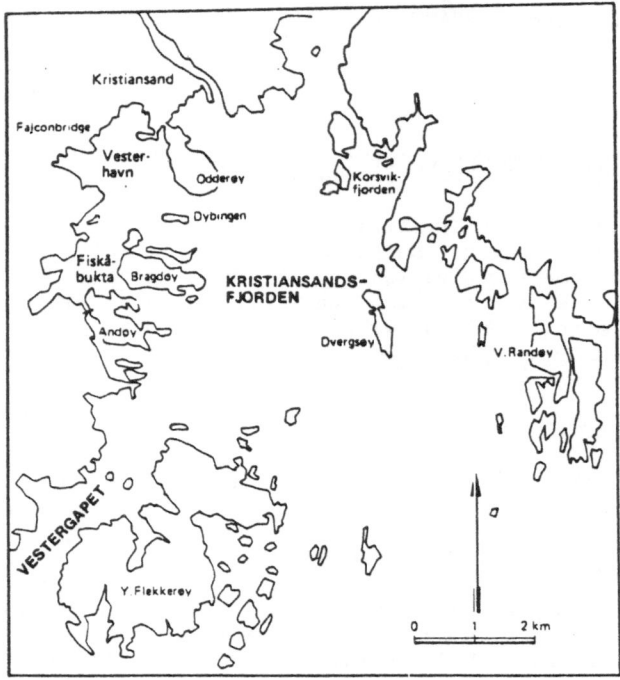

The map shows the central parts of the Kristiansand fjord.

A reduction in the concentrations of pollutants will reduce the health risk posed by eating fish and shellfish, and will increase the opportunities for leisure fishing and shell picking in the area. To achieve this there must be a reduction in the amount of organic micropollutants allowed to drain into the fjord.

C Less damage to soft bottom fauna in Vesterhavn and Fiskåbukta

Today the fauna in large areas of Vesterhavn-Fiskåbukta shows considerable damage. The number of organisms has dropped, and the diversity of species has also changed. Near Falconbridge Nikkelverk the soft bottom fauna has been completely destroyed. The principle cause of this damage is the present and earlier discharge of metals and metal-containing sludge (nickel, iron, copper, etc.) from Falconbridge.

Although the effluents have been substantially reduced since 1981-82, calculations show that parts of the area will continue to be affected if the effluents are allowed to remain at today's level. By further reducing the discharge of metals, the fauna in the fjord will return to normal within a time period of 30 to 50 years.

EVALUATION OF THE BENEFITS

Let us now look at measures that will bring the following benefits:

A) Water clarity in the whole of Vesterhavn and Fiskåbukta is improved to a level satisfactory for recreational purposes.

B) Fishing becomes possible in the whole Kristiansand fjord area.

C) The soft bottom fauna returns to normal.

212

1 NOW PUT THESE EFFECTS IN ORDER OF IMPORTANCE.
   Please note! We are interested in <u>your personal opinion</u>!

   A Clear water   ☐

   B Recreational fishing   ☐

   C Soft bottom fauna   ☐

   > Use 1, 2 or 3
   > 1 - most important
   > 2 - next most important
   > 3 - least important

2 NOW GIVE THE LEAST IMPORTANT EFFECT A VALUE OF 10.
   Try to give the other two effects a value according to how important
   you think they are in relation to the least important. Use the numbers
   10, 20, 30, and so on. For example, a value of 30 will mean the effect
   is 3 times as important as the one you considered least important. (If
   you think one of the effects is only as important as the least
   important, you may mark the box under 10 for that one too.)

   MARK THE TABLE BELOW
   Put one cross in each of the lines 2 and 3

Effect	10	20	30	40	50	60	70	80	90	100...*)
The least important effect	x									
The second most important effect	☐	☐	☐	☐	☐	☐	☐	☐	☐	☐
The most important effect	☐	☐	☐	☐	☐	☐	☐	☐	☐	☐

*) You can put a number here if you exceed 100.

BALANCING COSTS AND BENEFITS AGAINST EACH OTHER

In order to carry out a cleanup operation, the authorities must find a way
to finance such a project. How these costs will be distributed will depend
on the type of measures used and on who will carry them out.

Below two different alternatives are presented - try to evaluate them. For
each alternative try to weigh the cost against the benefit.

ALTERNATIVE I

Imagine that the discharge of sewage into Vesterhavn and Fiskåbukta is
eliminated by diverting the municipal sewage from the western part of
Kristiansand to Vestergapet. This will improve the water clarity to an
acceptable level all over Vesterhavn and Fiskåbukta. The water will become
as clear as in the outer parts of the fjord, and most areas in Vesterhavn
and Fiskåbukta can, after a while, be used again for bathing and similar
purposes.

These effects are the same as those called effect A above. Observe that this measure will not affect recreational fishing or the soft bottom fauna (effects B and C) to any appreciable extent.

The costs of this alternative are estimated to be about 55 million Norwegian Kroner (NOK). Most of this must be paid by the Kristiansand municipal area and its inhabitants. Each household will have to pay about 300 NOK annually.

3 I THINK (put one cross)

The benefits of this alternative will exceed the costs
        (Go on to question 4)   ☐

The costs of this alternative will exceed the benefits
        (Go on to question 5)   ☐

4 THE BENEFITS SEEM GREATER THAN THE COSTS (Alternative I)

Assign a value of 10 to the costs. Try to give the beneficial effects a value by how much larger they appear to be, compared with the costs.

    If, for example, you think it would be justifiable to spend triple the amount (165 million NOK) to achieve the effects just described, but not four times as much, place a cross in the box under 30. (If you think costs and benefits are about equal, mark the box under 10.)

Put one cross   10 20 30 40 50 60 70 80 90 100...*)

The costs of this alternative   ☒

The benefits of this
alternative   ☐ ☐ ☐ ☐ ☐ ☐ ☐ ☐ ☐ ☐

*) You can put a number here if you exceed 100.

5 THE COSTS SEEM GREATER THAN THE BENEFITS (Alternative I)

Assign a value of 10 to the benefits. Try to give the costs a value by how much larger they appear to be, compared to the benefits.

    If, for example, you think that one third of the amount (about 18 million NOK) is too much, but that one fourth is about right, mark the box under 40. (If you think costs and benefits are about equal, put a cross in the box under 10 for costs also.)

Put one cross   10 20 30 40 50 60 70 80 90 100...*)

The benefits of this
alternative   ☒

The costs of this alternative   ☐ ☐ ☐ ☐ ☐ ☐ ☐ ☐ ☐

*) You can put a number here if you exceed 100.

ALTERNATIVE II

In this alternative we will also carry out a total renovation of the
sewage system on the west side of the fjord, as in Alternative I. In
addition, the whole fjord will become suitable for fishing (effect B) and
the soft bottom fauna will again become normal in the greater part of
Vesterhavn-Fiskåbukta (effect C).

To obtain these effects the discharge of heavy metals and organic micropollutants from Falconbridge must be drastically reduced.

It is no simple matter to estimate the costs of these measures, but the
total cost, including Alternative I, will not exceed 200 million NOK. This
will have to be paid partly by the municipal authorities, partly by
Falconbridge, and partly by governmental support. Each household will have
to pay substantially more under this alternative than under Alternative I.

---

6 | I THINK (put <u>one</u> cross)

The benefits of this alternative will exceed the costs ☐
    (Go on to question 7)

The costs of this alternative will exceed the benefits ☐
    (Go on to question 8)

---

7 THE BENEFITS SEEM GREATER THAN THE COSTS (Alternative II)

Assign a value of <u>10</u> to the costs. In the same way as in alternative I,
try to give the benefits a value according to how great you think they
appear to be in relation to the costs. (If you think the costs and
benefits are about equal, give the benefits also a value of 10.)

Put one cross	10 20 30 40 50 60 70 80 90 100...*)
Costs	[x]
Benefits	☐ ☐ ☐ ☐ ☐ ☐ ☐ ☐ ☐ ☐

*) Put a number here if you exceed 100.

---

8 THE COSTS SEEM GREATER THAN THE BENEFITS (Alternative II)

Assign a value of <u>10</u> to the benefits. As in alternative I, give the
costs a value according to how great they seem to be in relation to
the benefits. (If you think the costs and benefits are about equal,
give the costs also a value of 10.)

Put one cross	10 20 30 40 50 60 70 80 90 100...*)
Benefits	[x]
Costs	☐ ☐ ☐ ☐ ☐ ☐ ☐ ☐ ☐ ☐

*) Put a number here if you exceed 100.

215

EVALUATION OF ALTERNATIVE I VERSUS ALTERNATIVE II

In this final question we would like you to evaluate the two alternatives against each other.

Under Alternative I the discharge of municipal sewage to the western parts of the fjord was eliminated, at a cost of about 55 million NOK (1986 price level). The water will appear much cleaner in Vesterhavn-Fiskåbukta as a result (effect A).

Alternative II comprised several measures and involved a more drastic reduction in the input of pollutants to the fjord. In addition to clearer water in Vesterhavn-Fiskåbukta, there will be an improvement in recreational fishing in the fjord (effect B) and the soft bottom fauna will develop normally in Vesterhavn-Fiskåbukta (effect C). Alternative II will cost approximately 200 million NOK, in other words a good deal more than Alternative I.

9   Give the alternative that you think is least desirable a value of 10. Then give the other alternative a value according to how many times better you think it is. Consider both the benefits and the costs when comparing these alternatives.

MARK THE TABLE BELOW
Put one cross in each of the lines 2 and 3

Evaluation of alternative I versus alternative II	10	20	30	40	50	60	70	80	90	100...*)
Alternative I	☐	☐	☐	☐	☐	☐	☐	☐	☐	☐
Alternative II	☐	☐	☐	☐	☐	☐	☐	☐	☐	☐

*) Put a number here if you exceed 100.

BACKGROUND INFORMATION

In a survey of this kind it is very important to present and analyze the results properly. To be able to do this we have to know a little about your background. This is solely to make proper statistics, and the information will be kept strictly confidential.

10 AGE: ____

11 INCOME: My gross annual income is about _____ NOK

12 EDUCATION:

        Primary/secondary school   ☐

        Further education   ☐

        University   ☐

13 OCCUPATION:

Not employed outside the home

        Housework in the home   ☐

        Pensioner/receiving social   ☐
        security

Employment

        Self employed   ☐

        Employed by others   ☐

        Agriculture, forestry,   ☐
        fishing, hunting

APPENDIX G: Summary report of the Pilot Study on
            Risk Management of Chemicals in the Environment[*]

1   INTRODUCTION

Public awareness of the risks posed by man-made chemicals to human
health and the environment has been increasing in recent years. We are not
only better informed about the dangers, but we have also seen dramatic
effects of accidential releases of hazardous chemicals into the
environment. In some cases, such as the accident in the nuclear reactor in
Chernobyl and the release of deadly methyl isocyanate from the chemical
plant in Bhopal, the adverse effects are immediately apparent. In other
cases, however, the adverse effects of chemicals may only become apparent
first after a long period of time as illustrated by the recent forest
damage in Central Europe and the decrease in stratospheric ozone in the
Antarctic. Under such circumstances it may be very difficult to establish
causal relationships with certainty. Irrespective of the existing
uncertainties, decisions on environmental problems must be taken. (To do
nothing is also a decision which may have serious consequences.) The
decision-makers are therefore in a very difficult situation. On the basis
of a huge amount of data, which may be interpreted in conflicting ways by
experts, they must try to reach a rational decision. To assist the
decision-makers various formal methods have been developed. So far all the
methods have considerable weaknesses. Often the process of applying the
methods is more useful than the results obtained.

2   PILOT STUDY DEVELOPMENT

Realizing the importance of improving environmental decision making,
Dr. H. C. Christensen, the Norwegian delegate to the 1983 Fall Plenary of
the Committee on the Challenges of Modern Society (CCMS), proposed a Pilot
Study on Risk Management of Chemicals in the Environment. A draft proposal
of the study, which had been written by Dr. Kari Kveseth, was presented.
Dr. Christensen informed the participants at the Plenary Meeting that
Norway was willing to act as the Pilot Country. The following objectives
were suggested:

- to review and develop methods through studies of specific examples
  related to chemicals in the environment,
- to identify areas for further research and development; and
- to establish methods for risk assessment and risk management that may
  be adopted by member countries.

An "expert meeting" with participants from Denmark, Greece, France,
and Italy, in addition to representatives from a number of Norwegian
institutions, was held in Oslo in April 1984. The attendees concluded that
a Pilot Study, as delineated in the draft proposal, would be useful.

The progress of the Pilot Study was initially not very promising.
Despite the general interest for the study, it proved difficult to get
people actively involved. Concern about the lack of progress was expressed
at the CCMS Plenary in November 1984. At that time it had not yet been
possible to work out a detailed programme for the study.

---

[*] Presented for the CCMS at the committee's Spring Plenary 1988

A second Pilot Study meeting was arranged in Oslo in March 1985. The Netherlands, represented by the TNO, and the USA, represented by the United States Environmental Protection Agency (EPA), had then decided to take part in the study. Unfortunately, EPA was not able to attend the second meeting. The participating countries were Belgium, Denmark, the Netherlands, and Norway.

Slowly the study gained momentum. This was due in part to active contributions from scientists receiving CCMS fellowships for working on the project. (C.F. Cavelli, Italy; C. Lupi, Italy; J.L. Regens, USA; D. Rondia, Belgium; C.L. van Deelen, the Netherlands). Further meetings were arranged in Washington (November 1985), Rome (April 1986), and in Liège, Belgium (November 1986). At the meeting in Liège the participants agreed on the content of the final report. A final meeting, at which the contributions to the final report were discussed, took place in Oslo in October 1987.

3   STRUCTURE AND OUTCOME OF THE PILOT STUDY

Although the Pilot Study was initially aimed at risk management, it soon became apparent that it was necessary to include aspects of risk assessment (hazard identification, dose/response assessment, exposure assessment, risk characterization). It should be to noted that while risk assessment in principle can be carried out objectively, risk management, which includes the process of weighing policy alternatives, necessarily must involve preferences and attitudes.

It is, of course, not possible to cover all aspects of risk assessment and risk management in a study of this kind. The participants were not chosen with this goal in mind. Nevertheless, many important topics were covered in the study. In general, the participants contributed projects in which they were involved when the Pilot Study was initiated. Each subproject was, therefore, primarily carried out as a national undertaking. However, the Pilot Study meetings and visits between Study Group members had in some cases considerable influence on the project development.

The Pilot Study resulted in useful cooperation among groups in the participating countries. The main outcome of the study is the report to be published by Plenum Publishing Corporation in the NATO CMS Series. The table of content for that report is provided in an appendix and is briefly summarized here.

Chapters 2 through 6 deal, to a large extent, with methods for risk assessment and risk management, though care has been taken to illustrate the methods by practical examples. Methods used in the United States for dealing with risks from chemicals to health and ecosystems are described in Chapters 4 and 5, while the attitudes of EPA employees toward risk assessment, cost-benefit analysis and the use of monetary values for human life are discussed in Chapter 6. Examples of dose-response relationships and related uncertainties for health effects are given in Chapters 2, 4, and 9. Even if there are numerous difficulties in assessing dose-response relations for health effects, it should be realized that the problems are still much greater with respect to ecosystem responses to chemicals (Chapters 5 and 10).

An example of a model to describe the fate of pollutants in soil is given in Chapter 7. Models to describe the fate of pollutants in water are briefly mentioned in Chapter 10.

Chapters 8 through 11 describe case studies. The first two are mainly on risk <u>assessment</u>: the Seveso accident in Italy (Chapter 8) and uncertainties in assessing acceptable concentrations of lead in blood (Chapter 9). The latter two chapters are mainly on risk <u>management</u>: valuation of countermeasures to pollution in a Norwegian fjord (Chapter 10) and to air pollution in Oslo (Chapter 11).

In most cases decisions are arrived at without full use of formal methods, either because of limited resources or limited confidence in such methods. The use of some simplified approaches is discussed in Chapters 3 and 11. The final chapter (Chapter 12) gives conclusions and recommendations to the CCMS; these are also included in Section 4 below.

4   CONCLUSIONS AND RECOMMENDATIONS

It can be concluded that formal methods are useful in risk assessment and risk management (e.g., cost-benefit analysis). Formal methods help to structure the problem, provide excellent opportunities for communication between decision-makers and groups affected by the problem, and may result in a ranking of decision alternatives. Such methods now seem to be gaining increasing acceptance by decision-makers.

In recent years considerable experience has been obtained in assessing and managing health risks. However, there is a serious lack of data in most cases both on exposure to chemicals and on the relation between dose and effects. This complicates the use of quantitative methods in risk assessment and risk management. These problems are still much more pronounced for effects of chemicals on ecosystems, especially for long-term exposure to fairly low doses.

In several countries attempts have been made to standardize approaches for assessing risks to human health and ecosystems, for example, by developing guidelines. Such guidelines may be useful; however, at this stage of development flexibility in choice of approaches to risk assessment and risk management is essential. The Pilot Study group, therefore, will not give specific recommendations regarding which methods to use and has made no attempt to develop detailed guidelines. However, two points should be emphasized: First, detailed risk analysis is not required under all circumstances. In order to assist the decision-maker simplified approaches may be sufficient. Second, it should be a general requirement that the uncertainties accumulating from all steps in the analysis appear clearly in the presentation of the results.

There are a large number of research needs related to the risk assessment and risk management of chemicals in the environment. In particular, we will mention the following areas:

-    An effort is needed to increase our understanding of harmful effects of chemicals on ecosystems.

-    There is a need for improved models for chemical fate in assessing risks both to health and the environment.

-    Exposure assessments are often highly uncertain and better methods and more data are required.

-    Better dose-response relationships are required in many cases, in particular for long-term effects.

- Valuation of effects will often include commodities for which there are no market prices. Continued research on how to elicit the public's preferences is of great importance. Emphasis should be placed on establishing methods that are comprehensible and easy to use and at the same time represent the public's preferences in an adequate way.

- New ways to include the affected groups in the decision process should be tried.

Specific recommendations to CCMS:

The committee is invited to:

- Work for increased efforts to protect humans and the ecosystem from harmful pollutants.

- Encourage membership countries to reduce emissions of harmful chemicals to the environment.

- Encourage international cooperation and interdisciplinary research in risk assessment and risk management. One way to achieve this is by continuing to provide CCMS fellowships and by giving collaboratory research grants in this field.

- Encourage national and international agencies to support work in this field and to use risk assessment and risk management methods in their work.

- Provide for means to follow up this Pilot Study with additional studies in related areas, e.g., studies of dose-response relationships for effects of chemicals on ecosystems or methods for eliciting the public's preferences.

- Contribute to improved information about the effects of chemicals on human health and the environment to the general public.

Appendix

Contents of RISK MANAGEMENT OF CHEMICALS IN THE ENVIRONMENT
(to be published by Plenum Publishing Corporation)

Chap. 1   Background
          H.M. Seip and A. Heiberg (Center for Industrial Research, Oslo,
          Norway)

Chap. 2   Quantification of Health Risk due to Chemicals: Methods and
          Uncertainties
          D.H. Trønnes and A. Heiberg (Center for Industrial Research,
          Oslo, Norway)

Chap. 3   Methods for Assessing the Risk of Environmental Contamination
          C.L. van Deelen (Netherlands Organization for Applied Scientific
          Research, Apeldoorn, the Netherlands)